Daniel Feurstein

INDIVIDUAL AND COMBINED NEUROTOXIC EFFECTS OF CYANOBACTERIAL TOXINS

Daniel Feurstein

INDIVIDUAL AND COMBINED NEUROTOXIC EFFECTS OF CYANOBACTERIAL TOXINS

Neurotoxic potential of Microcystins

Südwestdeutscher Verlag für Hochschulschriften

Imprint
Any brand names and product names mentioned in this book are subject to trademark, brand or patent protection and are trademarks or registered trademarks of their respective holders. The use of brand names, product names, common names, trade names, product descriptions etc. even without a particular marking in this work is in no way to be construed to mean that such names may be regarded as unrestricted in respect of trademark and brand protection legislation and could thus be used by anyone.

Publisher:
Südwestdeutscher Verlag für Hochschulschriften
is a trademark of
Dodo Books Indian Ocean Ltd., member of the OmniScriptum S.R.L Publishing group
str. A.Russo 15, of. 61, Chisinau-2068, Republic of Moldova Europe
Printed at: see last page
ISBN: 978-3-8381-2359-2

Zugl. / Approved by: Konstanz, Universitity of Konstanz, Germany, Dissertation, 2009

Copyright © Daniel Feurstein
Copyright © 2011 Dodo Books Indian Ocean Ltd., member of the OmniScriptum S.R.L Publishing group

Phantasie ist wichtiger als Wissen, denn Wissen ist begrenzt.

Albert Einstein (1879-1955)

Diese Dissertation ist unserem Vater gewidmet.

TABLE OF CONTENTS

ZUSAMMENFASSUNG 5
SUMMARY 7
1. CYANOBACTERIA – A GENERAL INTRODUCTION 9
   1.1. PHYSIOLOGY, ECOLOGY AND MASS OCCURRENCE OF CYANOBACTERIA 9
   1.2. HISTORY OF TOXIC CYANOBACTERIA 15
   1.3. CYANOBACTERIAL TOXINS 16
      1.3.1 Alkaloids and special toxin types 18
      1.3.2 Cyclic peptides 22
   1.4. MICROCYSTIN POISONINGS – A RISK TO PUBLIC HEALTH 31
   1.5. ORGANIC ANION TRANSPORTING POLYPEPTIDES 35
   1.6. INITIAL EVIDENCE FOR THE NEUROTOXIC POTENTIAL OF MICROCYSTINS – ....
        HISTORY AND GOAL OF THE STUDY 38
2. MANUSCRIPT I 41
   ABSTRACT 41
   INTRODUCTION 42
   MATERIALS AND METHODS 45
   RESULTS 51
   DISCUSSION 57
   ACKNOWLEDGEMENTS 59
3. MANUSCRIPT II 60
   ABSTRACT 60
   INTRODUCTION 61
   MATERIALS AND METHODS 62
   RESULTS 66
   DISCUSSION 71
   CONCLUSION 73
   ACKNOWLEDGEMENTS 73
4. MANUSCRIPT III 74
   ABSTRACT 74
   INTRODUCTION 75
   MATERIALS AND METHODS 76
   RESULTS 79
   DISCUSSION 87
   ACKNOWLEDGEMENTS 90
5. ADDITIONAL DATA 91
   5.1. MC IN VIVO STUDY 91
   5.2. PRELIMINARY RESULTS AND DISCUSSION 94
6. OVERALL DISCUSSION 98
   6.1. TOXIC CYANOBACTERIA: A GROWING RISK? 98
   6.2. MICROCYSTINS – POTENT NEUROTOXINS? 98
   6.3. INVOLVEMENT OF ORGANIC ANION TRANSPORTING POLYPEPTIDES IN MICROCYSTIN UPTAKE 100
   6.4. MICROCYSTIN CONGENER-DEPENDENT AND NEURON-SPECIFIC TOXICITY 102
   6.5. ASSESSMENT OF HUMAN HEALTH RISK, ARISING FROM MICROCYSTIN CONGENER DIFFERENCES AND ....
        THEIR NEUROTOXIC POTENTIAL 103
   6.6. CONCLUSIONS AND FUTURE PERSPECTIVES 104
7. REFERENCES 105
APPENDICES 119
   ABBREVIATIONS 119
   DANKSAGUNG 121

Zusammenfassung

Wasserverunreinigungen durch toxische Cyanobakterien sind ein weltweit wachsendes Problem, das ein ernstes Risiko für Menschen darstellt. Am wahrscheinlichsten sind humane Vergiftungen möglich durch chronische Vergiftungen mit niedrigen Toxinkonzentrationen, zum Beispiel durch Kontaminationen von Trinkwasser, Nahrung (Fisch, Garnelen, Nahrungsergänzungsmittel auf Blau-Grün Algen Basis) oder durch Freizeitaktivitäten in Gewässern, die ein Massenvorkommen von toxischen Cyanobakterien aufweisen. Microcystine (MC) repräsentieren die am häufigsten vorkommende Gruppe von zyklischen Heptapeptiden mit mehr als 80 verschiedenen strukturellen Varianten (Kongenere). Es konnte gezeigt werden, dass MC-LR im Vergleich zu anderen Kongeneren sowohl *in vitro* als auch *in vivo* toxisch ist, wobei momentan vermutet wird, dass der Mechanismus der Toxizität hauptsächlich von der spezifischen und irreversiblen Inhibition von Serin/Threonin Protein Phosphatasen (PP) herrührt. Durch ihre chemische Struktur und Größe ist es MC nicht möglich, durch Zellmembranen hindurch zu diffundieren, sondern sie benötigen hierfür Transporter, sogenannte organic anion transporting polypeptides (Nager Oatp / Mensch OATP). Des Weiteren besitzen nicht alle Oatps/OATPs dieselbe Affinität und Kapazität für all diejenigen MC Kongenere, die während einer Cyanoblüte auftreten können. Deshalb ist die Verteilung, d.h. die Toxikokinetik von einzelnen bzw. mehreren Kongeneren, absolut abhängig von der spezifischen Expression von Oatps/OATPs im Gewebe und in dessen Zellen. Erste Anhaltspunkte nach humanen und tierischen (Wild- und Haustiere) MC Intoxikationen deuten auf eine Hepato-, Nephro- und Neurotoxizität hin. Tatsächlich findet man Oatps/OATPs in Hepatozyten und kortikalen Epithelzellen der Niere, die auch einzelne MC Kongenere transportieren können. Um eine Neurotoxizität induzieren zu können, sollten daher Oatps/OATPs nicht nur in der Blut-Hirn-Schranke, sondern auch in neuronalen Membranen exprimiert werden.

Ziel dieser Arbeit war es zu untersuchen, ob MC potente Neurotoxine darstellen. Hauptaugenmerk wurde auf eine Neuronen spezifische sowie MC Kongener abhängige Toxizität gelegt. In ersten *in vitro* Versuchen mit primären murinen Whole Brain Cells (mWBCs) sollten Anhaltspunkte gesammelt werden, die auf eine generelle MC Kongener abhängige Neurotoxizität hindeuten. Für eine genauere Abschätzung sollte in weiteren Versuchen mit primären murinen Cerebellar Granule Neurons (mCGNs), d.h. primäre Neurone, die Präsenz von mOatps, ein MC Kongener abhängiger Transport, PP Inhibition sowie eine in Folge auftretende Neuritendegradation und Cytotoxizität (Nekrose/Apoptose) untersucht werden. Abschließend sollte ein erster *in vivo* Versuch mit Mäusen die beobachteten *in vitro* Effekte bestätigen.

Exposition von mWBCs mit einzelnen MC Kongeneren führte zu einer Konzentrations- und MC Kongener abhängigen Cytotoxizität, wobei MC-LF die toxischste Variante darstellte, gefolgt von

MC-LW und MC-LR. Die MC Kongener spezifische Toxizität zeigte sich in Assoziation mit einer mOatp abhängigen Aufnahme, da in Kompetitionsexperimenten mit den Oatp/OATP Substraten TC (Taurocholat) und BSP (Bromosulfophthalein) eine reduzierte Cytotoxizität gezeigt werden konnte. Da sich mWBCs aus unterschiedlichsten Zell-Typen (z.b. Neuronen, Astrocyten und Microglia) zusammensetzten, konnte eine Neuronen spezifische MC induzierte Toxizität nicht bestätigt werden. Deshalb wurden im weiteren Verlauf primäre mCGNs verwendet (mehr als 95% Neurone), um die Präsenz von mOatps und eine MC Kongener abhängige Aufnahme zu untersuchen. Auf mRNA Ebene waren alle mOatps mit Ausnahme von Mitgliedern der Oatp6 Familie detektierbar. In weiterer Folge konnte die Expression von mOatp1b2, einem bereits beschriebenen MC-LR Transporter, auf Proteinebene bestätigt werden. ^{3}H-TC und ^{3}H-Estron Sulfat (ES), zwei Oatp/OATP Substrate, zeigten Sättigungskinetiken in mCGNs und bestätigten somit die funktionelle Expression von mOatps. Co-Inkubationen mit ^{3}H-TC oder ^{3}H-ES und einzelnen MC Kongeneren zeigten eine reduzierte Aufnahme von ^{3}H-TC und ^{3}H-ES. Diese inhibierte Aufnahme variierte zwischen 20 und 45%, hauptsächlich in Abhängigkeit von dem eingesetzten MC Kongener. MC-LF war der stärkste Inhibitor bezogen auf die Aufnahme von ^{3}H-TC und ^{3}H-ES in mCGNs, gefolgt von MC-LW und MC-LR. Weitere Methoden wie Western-Blot und PP Inhibitionstest bestätigten die Aufnahme von MCs in mCGNs wie auch deren kovalente Bindung an cytosolische PP. Weiters induzierten sowohl MC-LF als auch MC-LW bereits bei niedrigen nicht cytotoxischen Konzentrationen eine Neuritendegradation, wobei für MC-LR höhere cytotoxische Konzentrationen notwendig waren um ähnliche Effekte zu erzielen. Es konnte gezeigt werden, dass die Neuritendegradation mit einer Hyperphosphorylierung des Tau-Proteins einher geht. Bei höheren MC Konzentrationen konnte eine Caspase-3/7 abhängige Apoptose beobachtet werden, allerdings nur in MC-LF und MC-LW exponierten Neuronen.

Zusammengefaßt zeigen die Daten eine mOatp abhängige Verteilung von MCs in mWBCs und mCGNs sowie einen MC Kongener anhängigen Cytotoxizitäts-Mechanismus, der MC-LF als den potentesten neurotoxischen MC Kongener darstellt. Eine Bestätigung der *in vitro* Daten war jedoch *in vivo* noch nicht möglich. Da einzelne MC Kongenere sehr stark in ihrem neurotoxischen Potential variierten, sollte die momentane Risikobewertung, die nur auf MC-LR beruht, neu überarbeitet werden. Dies ist von höchster Wichtigkeit, da MC-LF häufig in Oberflächengewässern einer auftretenden Cyanoblüte zu finden ist. Folglich zeigen die Daten dieser Studie das Vorkommen eines neuen bis jetzt noch nicht in Betracht gezogenen Risikos durch MC für Mensch und Tier.

Summary

Contamination of natural waters by toxic cyanobacteria is a growing worldwide problem, representing serious risks to public health. Human poisonings have been associated with chronic exposure to low toxin concentrations via drinking water, contaminated food (e.g. fish, prawns, Blue-Green Algae Supplements) or as a result of recreational water activities in surface waters with toxic cyanobacteria blooms. Microcystins (MCs) are the most commonly found group of cyclic heptapeptide cyanotoxins with more than 80 structural variants (congeners). MC-LR, in comparison to other MC congeners, was demonstrated to be toxic *in vitro* and *in vivo*, whereby the mechanism of toxicity (toxicodynamics) is currently assumed to primarily stem from the specific and irreversible inhibition of serine/threonine protein phosphatases (PPs). However, due to their structure and size, MCs can not penetrate cell membranes by simple diffusion but rather require organic anion transporting polypeptides (rodent Oatp / human OATP) for an active uptake. Moreover, not all Oatps/OATPs have an identical affinity and capacity for MC congeners present in an actual cyanobacterial bloom situation. Therefore, the distribution i.e. toxicokinetics of an individual MC congener(s) appear to entirely depend on the tissue and cell-type inherent expression of specific Oatp/OATP transporters. Initial evidence from human and animal (wild and domestic) MC intoxications suggests that MCs can elicit hepato-, nephro- as well as neurotoxicity. Indeed, the presence of Oatp/OATP in hepatocytes and renal cortical epithelial cells have been demonstrated and also shown to transport some of the MC congeners. Consequently, OATPs/Oatps should be present not only at the blood-brain barrier, but also within the neuronal membrane in order that MCs can induce the assumed neurotoxicity.

The overall goal of this work was to investigate whether or not MCs represent potent neurotoxins *in vitro* and *in vivo* with a focus on neuron-specific toxicity induced by three different MC congeners. During initial *in vitro* experiments using primary murine Whole Brain Cells (mWBCs) preliminary evidence for single MC congener dependent general neurotoxicity should be assessed. For a more refined assessment of neurotoxicity, primary murine Cerebellar Granule Neurons (mCGNs), i.e. primary neurons, should allow determination of the presence of mOatps, MC congener- dependent uptake, and ensuing PP inhibition, neurite degeneration and cytotoxicity (necrosis/apoptosis). Finally, a first *in vivo* experiment with mice should allow confirmation of the effects observed *in vitro*.

Exposure of mWBCs to single MC congeners resulted in a concentration- and MC congener dependent cytotoxicity with MC-LF being the most toxic congener followed by MC-LW and MC-LR. The observed MC congener specific toxicity appeared associated with a mOatp dependent uptake, as competition experiments with the Oatp/OATP substrates TC (taurocholate) and BSP (bromosulfophthalein) resulted in lower overall cytotoxicity. However, since mWBCs represent a mixture of different cell-types (e.g. neurons, astrocytes and microglia), neuron-specific MC mediated toxicity could not be confirmed. Consequently, primary mCGNs, representing >95% neuronal cells were used, which allowed specific screening for the presence of mOatps and the determination of MC congener- dependent uptake. All known mOatps were present at the mRNA level, except for members of the mOatp6 family. The expression of mOatp1b2, a known MC-LR transporter, could be further confirmed at the protein-level. ^{3}H-TC and ^{3}H-estrone-sulfate (ES), two Oatp/OATP substrates, followed saturation kinetics in mCGNs thereby confirming the functional expression of mOatps. Moreover, upon co-incubation of mCGNs with ^{3}H-TC or ^{3}H-ES and single MC congeners, uptake of ^{3}H-TC and ^{3}H-estrone-sulphate was inhibited. This uptake inhibition varied between 20 and 45%, largely depending on the MC congener used. Indeed, MC-LF presented with the strongest competition for ^{3}H-TC and ^{3}H-ES uptake into mCGNs, followed by MC-LW and -LR. Further analyses, Western-Blot (WB) and PP inhibition assays, confirmed uptake of MCs by mCGNs as well as the covalent binding of single MC congeners to cytosolic PPs. Moreover, MC-LF and MC-LW induced neurite degeneration was already observed at non-cytotoxic concentrations, whereas higher cytotoxic concentrations were required for MC-LR to induce a similar degeneration. The MC induced neurite degeneration was demonstrated in conjunction with hyperphosphorylation of the tau-protein. At higher MC concentrations a caspase-3/7 dependent apoptosis was observed in MC-LF- and MC-LW-exposed mCGNs, whereas this was not the case for MC-LR.

In conclusion, above results confirmed mOatps dependent distribution of MCs in mWBCs and mCGNs. Moreover, the MC congener dependent mechanism of cytotoxicity was demonstrated, pointing to MC-LF as being the most potent MC for potential neurotoxic effects. Unfortunately, the *in vivo* confirmation of these findings was not yet possible. As individual MC congeners differ strongly in their potential neurotoxicity, the current risk assessment, based solely on MC-LR, may need a revision. The latter is of utmost importance, as the most potent potential neurotoxic MC, MC-LF, occurs with regularly in cyanobacterial blooms of surface waters. Thus, the results of these studies suggest the presence of a new and not yet carefully considered hazard for humans and animals.

1. Cyanobacteria – a general introduction

Cited foreword of the World Health Organization book (WHO, 1999) Toxic Cyanobacteria in Water: A guide to their public health consequences, monitoring and management.
"Concern about the effects of cyanobacteria on human health has grown in many countries in recent years for a variety of reasons. These include cases of poisoning attributed to toxic cyanobacteria and awareness of contamination of water sources (especially lakes) resulting in increase cyanobacterial growth. Cyanobacteria also continue to attract attention in part because of well-published incidents of animal poisoning..." (WHO, 1999).

1.1. Physiology, ecology and mass occurrence of cyanobacteria

Cyanobacteria (also known as blue-greens, blue-green algae, myxophyceans, cyanophyceans, cyanophytes, cyanobacteria, cyanoprokaryotes, etc.) are one of the most diverse groups of gram-negative prokaryotes in terms of their morphology, physiology and metabolism (Codd, 1995).
Contrary to eukaryotes like real algae and plants, cyanobacteria have posses a cell wall structure composed of a peptidoglycan layer and lack a membrane-bound nucleus and membrane-bound sub-cellular organelles (Hoiczyk and Hansel, 2000). Furthermore, cyanobacteria contain 70S rather than 80S ribosomes (Gray and Herson, 1976).
Based on the International Code of Botanical Nomenclature the class cyanophyceae represents 150 genera that include 2000 cyanobacterial species (van den Hoek, 1995), with noticeable differences in their morphology and physiology (Castenholz and Waterbury, 1989). Thus, they are systematically classified into five orders:

- *Chroococcales* (e.g. *Microcystis, Radiocystis, Snowella*)
- *Nostocales* (e.g. *Anabaena, Aphanizomenon, Cylindrospermopsis*)
- *Oscillatoriales* (e.g. *Planktothrix, Lyngbya, Spirulina*)
- *Pleurocapsales* (e.g. *Xenococcus, Myxosarcina, Pleurocapsa*)
- *Stigonematales* (e.g. *Fischerella, Symphyonema*)

The basic morphology defines unicellular, colonial and multicellular filamentous forms (Beardall *et al.*, 2009). Unicellular species can occur as a single cell (e.g. *Synechococcus* sp.) or form aggregates of hundreds of cells (e.g. *Microcystis aeruginosa*). In addition, a wide variety of filamentous species are known to form heterocysts and akinetes (e.g. *Anabaena* sp., *Nostoc* sp.) although some do not (e.g. *Trichodesmium* sp.) (Stewart *et al.*, 1975; Adams and Carr, 1981; Carpenter *et al.*, 1992). Cyanobacteria reproduce asexually but the method can differ within the aforementioned

1. Cyanobacteria – a general introduction

morphological forms. Unicellular species undergo simple cell division, whereas filamentous forms reproduce by trichome fragmentation or by the formation of special reproductive segments of the trichome, the so called hormogonia (Mur *et al.*, 1999).

Cyanobacteria are also known as symbionts in a variety of organism, like the marine diatom *Rhizosolenia*, the feather moss *Pleurozium schreberi*, the hornwort *Leiosporoceros dussii*, the roots of *Cycas* and leaves of *Azolla* (Adams and Duggan, 2008).

Presently, cyanobacteria are involved in the formation of stromatolites and laminated biogenic rocks (Reid *et al.*, 2000). Micro fossils and carbon isotope data from these formations suggest the involvement of cyanobacteria already billion of years ago (Paerl *et al.*, 2001a; Neilan *et al.*, 2002). Additionally, fossil stromatolites have also been found from the early Archean to Precambrian time period, 3.5 to 0.5 billion years ago that contain cyanobacteria (Schopf and Packer, 1987; Sergeev *et al.*, 2002). Furthermore, it was suggested by Awramik (Awramik, 1992) that cyanobacterial photosynthesis was responsible for the oxygenation of the atmosphere approximately 3.5 billion years ago. Thus, cyanobacteria belong to one of the oldest life forms and have probably played a major role throughout the biological development of the Earth.

Among the oldest organisms, cyanobacteria have evolved to produce a variety of bioactive compounds, including alkaloids and peptides which induce threats to human and environmental health. They are often associated with economic damage, but equally demonstrate potential for the development of pharmaceuticals (e.g. antibiotics, anti cancer drugs) and other biological applications like algaecides, herbicides and insecticides (Berry *et al.*, 2008). Nevertheless, very little is known about the functional role of these secondary metabolites in the physiology, ecology and natural history of these organisms.

Most cyanobacteria are photoautotrophic prokaryotic organisms (Stanier and Cohen-Bazire, 1977; Castenholz and Waterbury, 1989). They represent a unique group since they are the only prokaryotes that exhibit a plant-like oxygenic photosynthesis in which two photosystems (PS II and PS I) are connected in series using pigments like chlorophyll-a, carotinoids and accessory pigments such as phycocyanin, allophycocyanin and phycoerythrin (phyobiliproteins) to harness light energy to convert inorganic carbon into organic compounds, while at the same time liberating oxygen (Boichenko, 2004). This allows cyanobacteria to use widely available sunlight as energy source, CO_2 as carbon source and H_2O as an electron donor for basic growth requirements.

Next to the oxygenic photosynthesis, some cyanobacteria are also known to occur under anoxic conditions using sulfide as electron donor (anoxygenic photosynthesis) instead of H_2O (Post and Arieli, 1997).

1. Cyanobacteria – a general introduction

Furthermore, a few other species assimilate organic compounds such as sugar in the light (photoheterotrophy) (Pelroy et al., 1972; van der Meer et al., 2003).
Presumably all cyanobacteria degrade glycogen in the dark via the oxidative pentose phosphate pathway. This pathway results in the complete oxidation of glycogen to CO_2 while energy is obtained by transferring electrons, generated during glycogen catabolism to oxygen, through aerobic respiration (Summers et al., 1995).
In addition, some cyanobacteria (e.g. *Microcystis aeruginosa*, *Nostoc* sp. *Oscillatoria limosa*) start fermentation of endogenous storage carbohydrates immediately upon transfer to dark anoxic conditions (Tamagnini et al., 2007) and therefore are capable of surviving for very long periods under these conditions. This is the case for *Microcystis aeruginosa*, which overwinters on lake-bottom sediments in complete darkness for 6-9 months of the year and additionally in intense packed surface scums of highly buoyant *Microcystis aeruginosa* (Reynolds et al., 1981).

However, anoxia, high UV exposure, high temperatures, and high levels of iron, sulfide and methane were all factors that influenced early life on Earth. A comparison of "modern" and Precambrian living forms confirms the inference that cyanobacteria communities are very conservative and have changed insignificantly both morphologically and physiologically during the past two billion years (Sergeev et al., 2002). This may explain why cyanobacteria can thrive under conditions of environmental stress and in extreme habitats where they are able to out-compete other organisms (Paul, 2008).
In many environments typical growth requirements (e.g. sunlight, CO_2 and water) are in ample supply. This is often not the case with nitrogen. In cyanobacteria, nitrogen is present in polypeptides, such as proteins, in the nucleic acids, in the cell wall as peptidoglycan and in chlorophyll. In addition, cyanobacteria can take up nitrogen in different combined forms such as nitrate, ammonium, organic nitrogen, urea or as molecular dinitrogen (N_2), the largest pool of nitrogen on Earth (Stewart, 1973; Dahlman et al., 2004; Flores et al., 2005). However, because of the very stable triple bond between the two nitrogen atoms, only certain specialized organisms are capable of using N_2 as a nitrogen source. For the N_2 catalysis into ammonium (NH_4), high amounts of energy are required (Flores et al., 2005). Although this presents a problem for most organisms in the natural environment, cyanobacteria as photoautotrophic organisms use light and water for this purpose (Stewart, 1973; Stewart et al., 1975).
In addition, the multimeric enzyme complex nitrogenase is fundamental for dinitrogen catalysis (Berman-Frank et al., 2003). Since nitrogenase is extremely sensitive to oxygen it can only be active in a totally anoxic environment (Fay, 1992). Its occurrence in cyanobacteria is paradoxical, since cyanobacteria not only usually live under aerobic conditions but also produce oxygen

1. Cyanobacteria – a general introduction

intracellularly as a product of photosynthesis (Fay, 1992). In general, three types of nitrogen fixing cyanobacteria can be distinguished. To overcome the aforementioned problems, cyanobacteria have developed two major strategies. First of all, the entrance of oxygen into the cell must be prevented or at least limited and additionally nitrogen fixation must be separated from oxygenic photosynthesis (Fay, 1992; Berman-Frank et al., 2003).

Heterocystous cyanobacteria (e.g. *Anabaena* sp., *Nostoc* sp., *Aphanizomenon* sp., *Nodularia* sp.) have solved this problem by differentiating special cells – the heterocysts (Stewart et al., 1975). The heterocyst has an extraordinary thick cell wall that presents a diffusion barrier for gases and demonstrates an anoxic environment by eliminating oxygen from the site of nitrogen fixation (Adams and Carr, 1981). Although the heterocyst can provide nitrogenase with ATP, reducing equivalents have to be imported from the vegetative cell, which in turn receive fixed nitrogen.

In addition, many heterocystous cyanobacteria can also form a second specialized cell type, called an akinete. Akinetes are thick-walled cells, representing a kind of spore (not heat resistant like bacterial endospores) with the ability to store reserve materials and to germinate under suitable growth condition (Adams and Carr, 1981; Mur et al., 1999).

A large number of filamentous and unicellular anaerobic N_2-fixing non-heterocystous cyanobacteria (e.g. *Plectonema boryanum*, *Oscillatoria limnetica*, *Synechococcus* sp.) are known to synthesize nitrogenase (Rippka et al., 1979). The number of reports of non-heterocystous cyanobacteria, unicellular as well as filamentous, that are capable of nitrogen fixation under fully oxic conditions while carrying out oxygenic photosynthesis is increasing (e.g. *Oscillatoria* sp., *Trichodesmium* sp., *Lyngbya* sp., *Microcoleus* sp.) (Fay, 1992; Sroga, 1997). The strategy by which such organisms protect nitrogenase against oxygen is not known precisely.

Two of the cyanobacterial key metabolic activities, namely oxygenic photosynthesis and dinitrogen fixation, are extremely sensitive to UV light-mediated injury (Garcia-Pichel, 1998). Avoidance of UV light is therefore important since it negatively affects photosynthesis of cyanobacteria and of course other phototrophic microorganisms (Garcia-Pichel et al., 1994). Nevertheless, especially benthic cyanobacteria exposed to high light irradiations, synthesize and accumulate UV light absorbing mycosporine-like amino acids (MAAs; e.g. colorless and water soluble shinorine, mycosporine-glycin, porphyra-334 and asterina 330) as well as the yellow-brown lipid-soluble pigment called scytonemin (Oren and Gunde-Cimerman, 2007; Paul, 2008).

With the accessory pigments phycocyanin, allophycocyanin and phycoerythrin, cyanobacteria are able to use effectively light between 550 and 650 nm wavelength, the light spectrum between the absorption peaks of chlorophyll-a and the carotinoids, which enables cyanobacteria to colonize a wide range of ecological niches. In addition, most cyanobacteria can undergo chromatic adaption,

1. Cyanobacteria – a general introduction

thus producing the accessory pigment needed to absorb light most efficiently in the environment in which they are present (Tandeau de Marsac, 1977; Mur et al., 1999).

Next to "sunscreen pigments" cyanobacteria have remarkable abilities to store essential nutrients and metabolites with known and unknown function like glycogen, lipid globules, cyanophycin, ploy-b-hydroxybutyrate (uncertain function), polyphosphate (Kenyon et al., 1972; Allen, 1984) and toxic secondary metabolites (Sivonen and Jones, 1999). Another special feature of cyanobacteria is the ability of regulating buoyancy (Oliver and Walsby, 1984; Oliver, 1994). Thus, they are able to vary their vertical position in the water column and subsequently allow an optimal exploitation of light and nutrient resources.

Furthermore, cyanobacteria often favor warm water temperatures for their growth (e.g. > 30°C for *Cylindrospermopsis raciborskii*) (Saker and Eaglesham, 1999) exceeding the optimal temperature of most green algae. *Phormidium corallyticum* for instance causes the common coral disease (black band disease) especially at water temperatures of 30-37°C (Cooney et al., 2002).

Special cyanobacteria features, as mentioned above like nitrogen fixation, regulated buoyancy, tolerance to UV light, high temperature tolerance, secondary metabolites, resting cells (e.g. akinetes), chromatic adaption as well as the specialized metabolism and their capacity to switch from one mode to another, make cyanobacteria extremely successful in a wide range of environments and very competitive to other microalgae. They can become the dominant primary producers, the base of the trophic food web, in intertidal areas, freshwater lakes and rivers, plant-free soils and paddy fields (Garcia-Pichel et al., 1996; Ward et al., 1998; Mur et al., 1999; Guven and Howard, 2006). This is especially true for environments that may be considered extreme like hypersaline pools, hot-springs, deserts sand, volcanic ash, rocks, alkali lakes and polar areas where no other microalgae can exist (Mur et al., 1999; Hitzfeld et al., 2000; Callaghan et al., 2004; Jungblut et al., 2005; Gorbushina, 2007; Miller et al., 2007). For example, in coral reef habitats cyanobacteria are becoming increasingly dominant on degraded reefs because of their ability to tolerate the environmental conditions associated with anthropogenic impacts and global climate change (Paerl et al., 2003; Paerl and Huisman, 2008; Paul, 2008; Myers and Richardson, 2009). However, freshwater habitats with diverse trophic states are the most prominent areas for cyanobacteria (Mur et al., 1999).

Cyanobacterial mass occurrence (often also called cyanobacterial bloom) is a complex phenomenon which is not completely understood. Some special environmental conditions as well as the above described adaptable physiology of cyanobacteria are suggested to encourage bloom formation. Additionally it is noteworthy to say that cyanobacterial blooms have negative effects in many aquatic ecosystems worldwide including Africa (e.g. Lake Victoria, small artificial lakes in the Kruger National Park), North America (e.g. Lake Erie; several small lakes in Michigan, New York, Ohio and Florida), Asia (e.g. Lake Taihu and Chaohu in China; Lake Suwa in Japan), Australia (e.g. Lake Alexandria) and Europe (e.g. Baltic Sea; several Portuguse waters, Lake Ijsselmeer in Netherlands, Finnish fresh and coastal waters, Lake Wannsee in Germany) (Sivonen *et al.*, 1990; Chen *et al.*, 2005; Dittmann and Wiegand, 2006; Xie *et al.*, 2007; Backer *et al.*, 2008; Paerl and Huisman, 2008; Oberholster *et al.*, 2009) including pristine oligotrophic pre-alpine and alpine lakes (e.g. Lake Ammersee, Germany; small alpine pasture lakes, Switzerland) (Mez *et al.*, 1996; Mez *et al.*, 1997; Naegeli *et al.*, 1997 2001; Mez, 1997a).

Cyanobacterial blooms have probably been present since living memory (see chapter 1.2). Nevertheless, it seems to be the case that these mass proliferation events occurred more often during the past decades (Ernst, 2001a; Carmichael, 2008).

Global climate change is occurring, which is already causing changes in terrestrial and marine ecosystems (Pearl and Huisman, 2008). Based on geological records, paleobiological evidence, and physiological and ecological studies, cyanobacteria seem likely to benefit from environmental changes associated with global warming (Paerl *et al.*, 2003; Paerl and Huisman, 2008; Paul, 2008). Higher temperatures (above 25°C) resulted in maximum cyanobacterial growth rates, which could explain the appearance of most cyanobacterial blooms during summer (Paerl and Huisman, 2008). In addition, next to the already mentioned facts, changing weather conditions (e.g. rainfall patterns, floods, hurricanes) and consequences of them influencing the stability and morphology of the water body may also trigger bloom formation (Paerl *et al.*, 2003; Paul, 2008).

Cyanobacterial mass occurrence was often detected in eutrophic lakes and near shore coastal waters, a consequence of nutrient over-enrichment by urban, agriculture and industrialization. Thus, it was suggested that blooming species require high phosphorus and nitrogen concentrations. However, cyanobacterial mass occurrence of nitrogen fixing ecostrategists like *Anabaena* sp., *Aphanizomenon* sp. and *Nodularia* sp. can often be related to periodic nitrogen limitation. It is suggested that this phenomenon is due to the fact that cyanobacteria have a high affinity for phosphorus and nitrogen with an optimum ratio of 1:10-16, respectively and the ability to store nutrients. Consequently, they can out-compete other photosynthetic organisms under limited phosphorus and nitrogen concentrations which require a ratio of 1:16-23, respectively.

Furthermore, cyanobacterial populations are very stable in their occurrence because the have only a few enemies (e.g. viruses, bacteria, only little grazing by copepods, daphnids and protozoa)

(DeMott et al., 1991; Jones et al., 1994a; Fialkowska and Pajdak-Stos, 1997; Fialkowska and Pajdak-Stos, 2002) and posses the ability to prevent sedimentation by buoyancy regulation.

1.2. History of toxic cyanobacteria

Contamination of natural waters by toxic cyanobacterial mass occurrence represents a growing worldwide problem, causing serious water pollution and health hazards for humans and livestock (Falconer, 1999; Vasconcelos, 1999; WHO, 1999; Briand et al., 2003; Dietrich and Hoeger, 2005; Falconer, 2005a; Paerl and Huisman, 2008).

Toxic planktonic and benthic blooms and associated poisonings have already been suggested in the palaeontology literature describing mass mortalities of deer, forest elephants, rhinoceros and cave lions at a Pleistocene lake in Neumark-Nord, Germany, 1.8 million – 11000 years ago (Stewart et al., 2008). Analyses of sediment extracts have revealed cyanobacteria-specific carotenoides and the presence of microcystins (similar UV spectrum compared with *Microcystis aeruginosa* extract) with an unspecified microcystin congener profile. More recently it was suggested by Hoeger (Hoeger, 2003) that the oldest documented observation was reported in the Old Testament: "…all the waters that were in the river were turned to blood. And the fish that were in the river died, and the river stank, and the Egyptians could not drink of the water of the river" (Exodus 7: 20 – 21). Millennia later, *Oscillatoria tenuis* and the cyanobacterial toxin microcystin were isolated from the River Nil (Brittain et al., 2000).

However, rather than more previous anecdotal reports, the first scientific publication for toxic cyanobacteria was published in *Nature* by George Francis in 1878 (Francis, 1878) describing a bloom of *Nodularia spumingena* in a freshwater lake in South Australia, called Lake Alexandria. Francis reported that the incidental consumption of the thick and pasty scum resulted in the rapid death of sheep, horses, dogs and pigs within periods of 1 to 24 hours. In the later part of the 19th century, the same lake demonstrated numerous bloom formations associated with cyanobacteria-related poisonings and several hundred deaths. Nevertheless during the same time, the local Aboriginal people were already sensitized and aware of the problems caused by Lake Alexandria (Codd et al., 1994).

After the first scientific report by George Francis, the portion of primary literature covering the topic "cyanobacteria secondary metabolites – the cyanotoxins", numbered about 850 (1878 to 1992). From 1992 to 1996 the number of primary papers and reviews on toxic cyanobacteria was over 900, which in term has exceeded the amount of publications produced over the entire 114-year period (1878 - 1992) (Carmichael, 1997).

Throughout the 20th century numerous cyanobacteria-related poisonings of wildlife and livestock in all continents have been described (Briand et al., 2003; Stewart et al., 2008) and indicated only in the U.S. 81 published articles on cyanobacterial outbreaks (1883–2003) (Carmichael, 2008) whereas it is supposed that the estimated number of undetected cases is much higher.

However, the first published report about human exposure (1931) to cyanobacterial toxins occurred in Charleston, West Virginia during a massive *Microcystis* bloom in the Ohio and Potomac Rivers. An inadequate treatment of the drinking water coming from these rivers was suggested to be the cause of intestinal illness in approximately 5000 to 8000 people (Tisdale, 1931). Additional human intoxication by contaminated drinking water or other sources of water (e.g. via hot steam / aerosol formation during bathing and sauna) were documented from North and South America (e.g. Florida and Brazil), Africa (South Africa), Europe (e.g. Finland, UK), Asia (e.g. China), Australia including Oceania (e.g. Guam) (Kuiper-Goodman et al., 1999; Dietrich et al., 2008).

Probably due to national and international monitoring programs and advanced analytical detection methods, a variety of new highly toxic secondary metabolites were identified within the last few decades (more than 600 peptides or peptidic metabolites from various taxa until 2005) (Welker and von Dohren, 2006), some of them exhibiting dramatic environmental and economic problems with direct or indirect consequences for public health. In these responses, The First International Conference on Toxic Cyanobacteria, "The Water Environment Algal Toxins and Health" was held in 1981 with the 8th such conference scheduled to take place in Turkey in 2010.

All of these anecdotal reports and scientific publications demonstrate that cyanobacterial blooms and cyanobacteria-related poisonings are no recent development but also indicate a growing risk for the future due to the fact that the total number of published toxic cyanobacteria events has exponentially increased within the last 30 years (Carmichael, 2008). This is a likely consequence of freshwater and nearshore coastal water nutrition in combination with global warming (Paerl et al., 2003; Paul, 2008; Paerl, 2008a). Therefore, the responses of cyanobacteria and the release of cyanobacterial toxins to changing environmental patterns are important subjects for future research (Paul, 2008).

1.3. Cyanobacterial toxins

As already mentioned, 2000 cyanobacterial species were identified (van den Hoek and Jahns, 2002) including approximately 40 genera known to be responsible for cyanobacteria toxin poisonings but the main representatives are *Anabaena* sp., *Cylindrospermopsis* sp., *Lyngbya* sp., *Microcystis* sp., *Oscillatoria* sp. (*Planktothrix* sp.) and *Nostoc* sp. (Carmichael, 2001a; van Apeldoorn et al., 2007). Furthermore, these toxic species have been observed to be capable of producing several different

1. Cyanobacteria – a general introduction

toxin types as well as different toxin congeners of one toxin group (e.g. microcystin-LR, -LW, -LF, -RR, -YR) (Dietrich et al., 2008). Thus, cyanotoxins represent a diverse group of natural toxins regarding the chemical structure but also differ a lot from the toxicological standpoint.

Healthy bloom populations in laboratory and field studies produce only small amounts of extracellular toxins (~10-20% of the total toxin pool) whereas the biggest toxin portion accumulates intracellular (Sivonen, 1990a; Lindholm and Meriluoto, 1991; Jones and Orr, 1994; Negri et al., 1997). However, contamination of the surrounded water seems only to be the case, if not exclusively, during cell senescence, cell death (e.g. by bacteria) or cell lysis after treatment with algaecides (e.g. copper sulphate) (Jones and Orr, 1994; Sivonen and Jones, 1999). The highest published concentrations of the cyanotoxins microcystin and anatoxin-a(S) per liter of water were up to 25.000 µg/l and 3.300 µg/l, respectively (Sivonen and Jones, 1999).

Since cyanobacteria are ubiquitously distributed it is not surprising that their toxic secondary metabolites can be detected worldwide (Sivonen and Jones, 1999) and indicate responsibility for almost all known cases of fresh and brackish water intoxication (van Apeldoorn et al., 2007). Cyanotoxins can be classified by (Sivonen and Jones, 1999):
- Mechanism of the toxicity (e.g. hepatotoxic, neurotoxic or dermatotoxic)
- Chemical Structure

A classification by the mechanism of toxicity is in my opinion not useful because some cyanotoxins are poorly investigated from the toxicological point of view and additionally some cyanotoxins indicate multi organ toxicity (e.g. microcystins demonstrate hepato-, nephro- and possibly neurotoxicity). Therefore, I have classified the cyanotoxins by their chemical structure into (Sivonen and Jones, 1999; Metcalf and Codd, 2004; Welker and von Dohren, 2006; van Apeldoorn et al., 2007):

- Alkaloids: anatoxin-a, anatoxin-a(S), saxitoxins, cylindrospermopsin, aplysiatoxins and lyngbyatoxin
- Others: LPS, β-methyl-amino-L-alanine, mueggelone

- Peptides: Cyclic peptides: microcystin, nodularins, cyanopeptolines, microviridins; anabaenopeptins, cyclamides; crytophycins, lyngbyabellin B
  Linear peptides: aeruginosins, microginins, microcolins, mirabimids, aeruginosinamide, barbamide)

However, the following section deals only with a brief summary of some prominent or unique toxin types like saxitoxins, anatoxins, β-methyl-amino-L-alanine (BMAA) and nodularins. Special emphasis is on the cyclic heptapeptide microcystin and a more in depth description regarding mode of action, cellular and organ toxicity, transport mechanisms and human health problems.

1.3.1 Alkaloids and special toxin types

Saxitoxins (paralytic shellfish poisoning (PSP) toxins)

A neurotoxin group of alkaloids, called saxitoxins were discovered early last century and possess a well established place in human history. Saxitoxin intoxications have resulted in many reports worldwide about morbidities and mortalities in humans and animals via contaminated drinking water and seafood (e.g. shellfish) causing the phenomenon known as paralytic shellfish poisoning (PSP) (Rodrigue *et al.*, 1990; Negri and Jones, 1995; Rapala *et al.*, 2004; Rapala *et al.*, 2005; Falconer, 2008). PSP has been reported in more than 1000 cases during the last century in North and Central America with 109 deaths (Kuiper-Goodman *et al.*, 1999).

Saxitoxin is the only marine toxin that is declared as a chemical weapon and internationally regulated by the Organization for Prohibition of Chemical Weapons (OPCW, Den Haag, Netherlands). It was listed in Schedule 1 of the Chemical Weapon Convention (CWC) together with sarin, ricin and mustard gas. Moreover, saxitoxin was ranked among the five most potent toxins, presented in the "deutsche Kriegswaffenliste" (BGB1. IS. 385, 1998) and in the briefing book on chemical weapons (Swedish Defence Research Agency (FOI) number 2, 2002).
Within the environment, saxitoxins are widely distributed and found in cyanobacteria (e.g. *Anabaena cirinalis, Cylindrospermopsis raciborskii, Aphanizomenon flos-aquae, Planktothrix* sp.), dinoflagellates (e.g. *Alexandrium tamarense, Gymnodinium catenatum*) or accumulated in clams, oysters, mussels and scallops (Sivonen and Jones, 1999; van Apeldoorn *et al.*, 2007; Llewellyn, 2006).
Saxitoxin possess a special tricyclic structure with hydropurine rings (Figure 1.1), representing a large group of approximately 24 carbamate alkaloid toxins which are either non-sulphated (saxitoxins, STX), singly sulphated (gonyautoxins, GTX) or doubly sulphated (C-toxins) (van Apeldoorn *et al.*, 2007).
All saxitoxins act in nerve axon membranes in the same way by inhibiting the voltage gated Na^+ channel and thereby preventing flow of Na^+ ions with consequent blocking of nervous transmission (Kuiper-Goodman *et al.*, 1999; Llewellyn, 2006; van Apeldoorn *et al.*, 2007). As a consequence in general, laboratory animals show typical signs of neurotoxicity including nervousness, jumping,

jerking, ataxia and convulsions. Finally paralysis of the respiratory muscles results in death within a few minutes (Briand et al., 2003). Those surviving the first 12 hours often make of full recovery (Llewellyn, 2006). LD_{50} (lethal dose resulting in 50% deaths) toxicity values for saxitoxin to vertebrates mainly differ on the administration route but also within animals. Orally exposed dogs revealed LD_{50} of 181 µg/kg and mice and LD_{50} of 263 µg/kg (Llewellyn, 2006).

There is no international tolerable daily intake (TDI) value for saxitoxin available but most countries accept a tolerance level of e.g. 80 µg $STX_{equivalents}$ / 100 g mussel meat (van Apeldoorn et al., 2007). However, the highest published concentration of saxitoxin from an Australian bloom sample was 3400 µg/g dry weight (dw) (Sivonen and Jones, 1999).

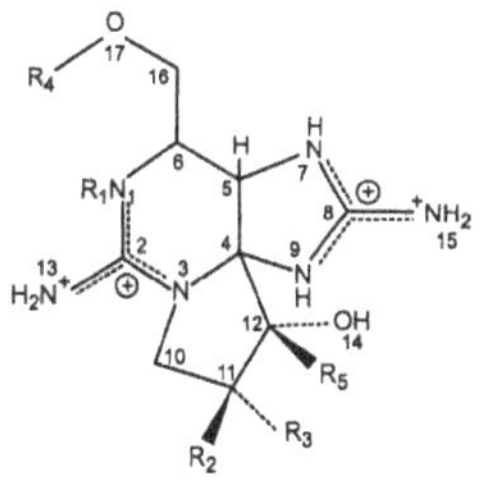

Figure 1.1: Chemical structure of PSP toxin
with R1=H, R2=H, R3=H, R4=CONH2, R5=OH for STX (Briand et al., 2003)

Anatoxin-a, and Homoanatoxin-a

The neurotoxic alkaloid anatoxin-a is known since the late 1950s as "very fast death factor" due to its high lethality (Carmichael et al., 1979; Osswald et al., 2007) within minutes. Many documented cases of animal fatalities, including cattle and dogs have been reported from several countries due to the consumption of contaminated water (Edwards et al., 1992; Osswald et al., 2007). A more recent case study from the south of France linked the neurotoxic symptoms of 37 dogs with 26 deaths (in 2002 and 2003) to an anatoxin-a intoxication (Stewart et al., 2008).

Several cyanobacteria strains producing anatoxin-a are described so far, including *Anabaena flosaquae*, *Anabaena cirinalis*, *Planktothrix* sp., *Aphanizomenon* sp, *Cylindrospermopsis* sp. and *Microcystis* sp. (van Apeldoorn et al., 2007).

Anatoxin-a is a tropane-related low molecular alkaloid (MW: 165 Da; Figure 1.2 A) with its homologue homoanatoxin-a (MW: 179 Da; Figure 1.2 B) isolated from e.g. *Oscillatoria formosa* (Sivonen and Jones, 1999). Homoanatoxin-a and anatoxin-a differ only at C-2, with a propionyl instead of an acetyl group for homoanatoxin-a (Skulberg et al., 1992).

Anatoxin-a is a potent postsynaptic blocking agent, targeting both but mainly nicotinic acetylcholine receptors, present in the peripheral nervous system (PNS) on the neuromuscular

1. Cyanobacteria – a general introduction

junction and to a lesser extent on muscarinic acetylcholine receptors, found in the PNS and central nervous system (Osswald et al., 2007). Therefore, anatoxin-a acts as a potent cholinergic agonist with a greater affinity for nicotinic acetylcholine receptors than the neurotransmitter acetylcholine (Osswald et al., 2007). As a consequence, cationic channels open, leading to a Na^+ influx and K^+ efflux, thereby depolarizing the neurons and initiating a new action potential. In a normal event, acetylcholine is hydrolyzed by the enzyme acetylcholinesterase which is not the case for anatoxin-a thus exhibiting an over stimulation. A sufficient dose can lead to a loss of muscle coordination, gasping, muscular paralysis and final death by asphyxiation (Carmichael and Biggs, 1978; Carmichael et al., 1979). The toxicity of homoanatoxin-a is due to an enhanced Ca^{2+} influx in the cholinergic nerve terminal (van Apeldoorn et al., 2007).

Acute toxicity studies revealed an i.p. LD_{50} of 375 µg/kg and an oral LD_{50} of greater than 5000 µg/kg for anatoxin-a (Fitzgeorge et al., 1994). Homoanatoxin-a demonstrated an i.p. LD_{50} of 250 µg/kg in mice leading to a respiratory arrest and subsequent death within 7 to 12 minutes (Kuiper-Goodman et al., 1999). In blooming samples from Finland an anatoxin-a concentration was discovered with 4400 µg/g dw (Sivonen and Jones, 1999). However, there is no official guidance value for anatoxin-a but 1 µg/l was suggested to represent an adequate margin of safety with regard to drinking water (Fawell et al., 1999a).

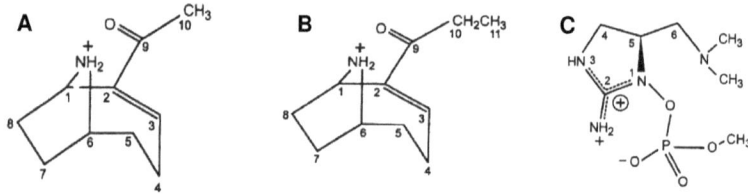

Figure 1.2: Chemical structures of Anatoxins
A: Anatoxin-a, B: Homoanatoxin-a, C: Anatoxin-a(S) (Briand et al., 2003)

Anatoxin-a(S)

Anatoxin-a(S) (Figure 1.2 C) is a very special natural toxin and represents similar toxicity to some chemical weapons. There are no human cases described so far associated with anatoxin-a(S) poisoning (van Apeldoorn et al., 2007; Humpage, 2008) but several deaths of dogs, pigs and ducks were reported in the USA (Briand et al., 2003) and waterbird deaths from Denmark (Stewart et al., 2008). The highest reported concentration of anatoxin-a(S) was found in blooming samples in the latter country with 3300 µg/g dw (Sivonen and Jones, 1999).

Only a few cyanobacteria species are described for producing anatoxin-a(S) e.g. *Anabaena flos-aquae* and *Anabaena lemmermannii* (Henriksen et al., 1997; Onodera et al., 1997).

1. Cyanobacteria – a general introduction

Anatoxin-a(S) is an organophosphate with a similar structure and mode of action like the chemical weapons sarin, soman, tabun an VX (briefing book on chemical weapons FOI number 2, 2002; personal communication E. van Elk, OPCW inspector) and organophosphate insecticides like Malathion and Parathion (Humpage, 2008).

Normally, achetylcholine is degraded by the enzyme achetylcholinesterase and thus silencing nervous action potential. In terms of an antitoxin-a(S) intoxication, achetylcholinesterase is blocked and resulted in similar effects to those of anatoxin-a but additionally with viscous mucoid hypersalvation and lacrimation (Falconer, 2008; Humpage, 2008).There are no oral toxicity data available (Falconer, 2008) but in acute toxicity studies an i.p. LD_{50} was demonstrated in mice and rats with 31 and 20 µg/kg, respectively. Because there are insufficient data, a TDI can not be calculated and there are not official regulations for anatoxin-a(S) (van Apeldoorn *et al.*, 2007).

BMAA

This very unique non-amino acid neurotoxin is likely the most controversial discussed cyanobacterial toxin (Duncan and Marini, 2006).

The BMAA hypothesis of Cox *et al* (Cox *et al.*, 2003) suggests that amyotrophic lateral sclerosis/Parkinsonism dementia complex (ALS/PDC), a common neurological disease among the Chamorro people of Guam, is caused by eating flying foxes who have BMAA accumulated, a consequence of their primary food source, namely cycad seeds. However, in a very interesting letter to the editor "Debating the Cause of a Neurological Disorder" Duncan and Marini (Duncan and Marini, 2006) have raised serious concerns about the BMAA hypothesis and their linkage to ALS/PDC. Finally they stated "The scientific community has been very receptive to the BMAA hypothesis; more than ever, the onus is now on its proponents to provide compelling and credible data." (Duncan and Marini, 2006). However, BMAA was detected in brain samples of 8 Chamorro people suffering from ALS/PDC and interestingly in 2 of 14 tested brain samples of Canadian Alzheimer's patients (Murch *et al.*, 2004).

BMAA was found in 95% of tested free-living cyanobacteria genera and in 97% of tested strains like *Microcystis* PCC 7806, *Aphanizomenon flos-aquae*, *Anabaena* PCC7120 and *Nostoc* PCC 6310, to name only some of them. In addition, BMAA was also isolated from symbiotic-living cyanobacteria like *Nostoc* PCC 7422 (host cycad) and *Nostoc* 8001 (host flowering plant) (Cox *et al.*, 2005). However, "it has been recently suggested that BMAA may be a new cyanobacterial toxin" (Humpage, 2008). That statement clearly demonstrates that, it is still not fully accepted in the scientific field that cyanobacteria are the main source of BMAA origin.

BMAA is structurally similar to methylated alanine (Figure 1.3) and once orally ingested it is able to cross membranes including the blood-brain-barrier via large neutral amino acid carrier (Duncan *et al.*, 1990; Duncan *et al.*, 1992). In the body, BMAA occurs in a "free" and protein-bound form. If BMAA is not bound to proteins it acts as an excitotoxin in glutamergic neurons by AMP/kainite receptor overstimulation. However, more recently protein-bound BMAA was suggested to function as an endogenous neurotoxic reservoir resulting in a slow release of the toxin to the cerebral tissue via protein metabolism. An incorporation of BMAA into proteins and subsequent misfolding seems to be likely but was not yet demonstrated. *In vivo* studies using monkeys dosed with high levels of BMAA (oral, 0.1 – 0.3 µg/kg day, up tp 12 weeks) developed similar effects to those observed in ALS/PDC patients (Spencer *et al.*, 1987) whereas two mice studies in which BMAA was administered orally in a chronic scenario (~0.5 µg/kg/day, over 11 weeks (Perry *et al.*, 1989)) revealed no neurotoxicity and no behavioral changes (Perry *et al.*, 1989; Cruz-Aguado *et al.*, 2006). "Much more work needs to be done before a proper assessment can be made of this "new" cyanotoxin" (Humpage, 2008).

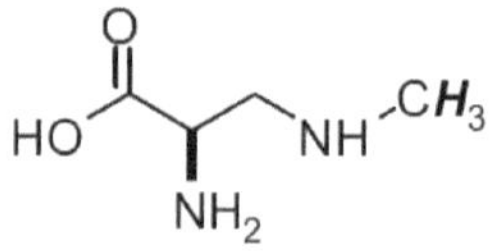

Figure 1.3: Structure of BMAA
(Moura *et al.*, 2009)

1.3.2 Cyclic peptides

Nodularin

Cyclic peptides of the nodularin and microcystin family are the most frequently detected cyanotoxins in fresh and brackish water blooms. In general, nodularins have similar properties as microcystins with regard to their chemical structure (Figure 1.4 for nodularin and Figure 1.5 for microcystins), mode of action and subsequent toxicity (Briand *et al.*, 2003; Humpage, 2008) but nodularin represents fewer cases of human and livestock poisonings (Kuiper-Goodman *et al.*, 1999). However, in two recent case studies intoxictation and subsequent death of several dogs were reported from Finland and South Africa (Persson *et al.*, 1984; Harding *et al.*, 1995). In the latter case, microscopic investigation indicated 95% of *Nodularia spumigena* in blooming samples and lyophilized material revealed a concentration of 3.5 mg/g dw nodularin. Although *Microcystis aeruginosa* was additionally present in the same bloom, microcystin was not detectable (Harding *et al.*, 1995).

So far, nodularins are only known to be produced by the species *Nodularia spumigena*, living in brackish water, for example in the Baltic Sea but was additionally isolated from Lake Alexandria, Australia and from New Zealand waters (Briand *et al.*, 2003; Humpage, 2008).

Nodularin is a cyclic pentapeptide (Figure 1.4) with a molecular weight of 824 Da (van Apeldoorn *et al.*, 2007), closely related to the structure of microcystins but with less structural variation (Wiegand and Pflugmacher, 2005; Humpage, 2008). The general structure of nodularins consists of cyclo-(D- erythro-b-methylaspartic acid L-arginine-Adda-D-glutamate-2-(methylamino)-2-dehydrobutyric acid), whereas Adda represents the unique amino acid (2S,3S,8S,9S)-3-amino-9-methoxy-2,6,8-trimethyl-10-phenyldeca-4,6-dienoic acid only found in nodularin and microcystin (van Apeldoorn *et al.*, 2007).

So far, seven structural variants have been reported via substitutions of arginine with homoarginine or by demethylation of e.g. D-erythro-b-methylaspartic acid (van Apeldoorn *et al.*, 2007; Humpage, 2008). One nodularin found in the marine sponge *Thenella swinhoei*, namely motuporin, consist of L-valine instead of L-arginine. However, it is suggested that motuporin is of cyanobacteria origin because the symbiotic-living sponge is known to host cyanobacteria (van Apeldoorn *et al.*, 2007).

Their mode of action, similar to microcystins, is due to the inhibition of serine/threonine (ser/thr)-specific protein phosphatases (PPs) especially PP1 and PP2A (Ohta *et al.*, 1994) but with the exception of covalent binding to the catalytic subunit of PPs (Bagu *et al.*, 1997). The latter observation is in conjunction with the smaller ring size of N-methyl-dehydrobutyrine instead of dehydroalanine and therefore prevents binding with the PP cysteine (Lanaras *et al.*, 1991; Craig *et al.*, 1996; Bagu *et al.*, 1997).

Acute toxicity studies in i.p. exposed mice revealed LD_{50}s of 30 to 50 µg/kg nodularin, causing death by liver hemorrhage or liver failure (van Apeldoorn *et al.*, 2007). Because of the absence of toxicological data, no non-observed-effect level (NOEL) can be derived and consequently no TDI.

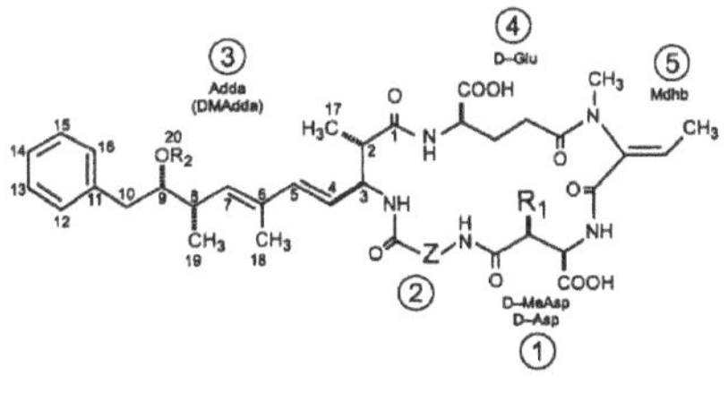

Figure 1.4: Structure of nodularins
(X and Z are variable amino acids) (Briand *et al.*, 2003)

1. Cyanobacteria – a general introduction

Microcystins (MCs)

The focus of this PhD thesis was restricted to three MC congeners. Therefore, a more detailed background is provided in the following chapters regarding the chemical structure of MCs, incidence, production and biodegradation, cellular uptake, mode of action, *in vitro* and *in vivo* toxicity and a summary of case reports indicating initial evidence for the neurotoxic potential of MCs.

The cyclic peptide group of MCs are globally the most widely distributed cyanotoxins in blooms of fresh and brackish waters and at the same time, most often involved in human and animal poisonings (Sivonen and Jones, 1999; Briand *et al.*, 2003; de Figueiredo *et al.*, 2004). The latter fact is especially true in conjunction with toxic blooms in water reservoirs, rivers, lakes and ponds used for drinking water or recreational purposes or as potable water sources by wild and domestic animals (Falconer and Humpage, 2005; Dietrich *et al.*, 2008). However, detailed information about human case reports, exposure routes and MC risk assessment are provided in Chapter 1.4.

MCs are produced non-ribosomally. Ten genes of the *myc*-cluster code for a mixed polyketide/peptide synthetase which represents a multi enzyme complex (Welker and von Dohren, 2006). Interestingly, *mycH* of the *myc*-cluster codes for an ABC transporter and therefore providing cyanobacteria with a mechanism to export their toxins in the environment (Welker and von Dohren, 2006).

MCs are small monocyclic peptides composed of seven peptide-linked amino acids, including an unique amino acid called Adda ((2S,3S,8S,9S)-3-amino-9-methoxy-2,6,8-trimethyl-10-phenyldeca-4,6-dienoic acid) (Zurawell *et al.*, 2005; van Apeldoorn *et al.*, 2007). Their chemical structure was first identified in the early 1980s, isolated from the cyanobacterium *Microcystis aeruginosa* and named after this organism (Carmichael *et al.*, 1988). The general structure is cyclo-(D-alanine-X-D-MeAsp-Z-Adda-D-glutamate-Mdha) in which X and Z are variable L amino acids, D-MeAsp is D-erythro-β-methylaspartic acid, and Mdha is N-methyldehydroalanine (Sivonen and Jones, 1999; Briand *et al.*, 2003; van Apeldoorn *et al.*, 2007). The presence of two variable amino acids (X and Z), two groups (R1 and R2; R = H or CH_3) and two demethylated positions (3, D-MeAsp and/or 7, Mdha) (Briand *et al.*, 2003; van Apeldoorn *et al.*, 2007) results in more than 80 structural variants (Spoof, 2005) with a size between 909-1115 Dalton (Da) (Sivonen and Jones, 1999; Zurawell *et al.*, 2005).

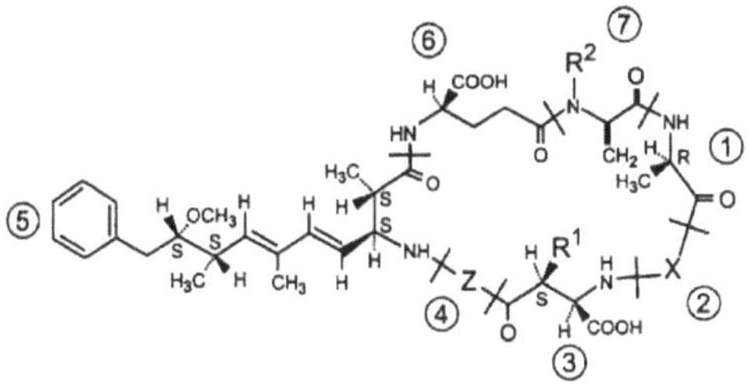

1. Cyanobacteria – a general introduction

Figure 1.5: Structure of microcystins
(X and Z are variable amono acids) (Briand et al., 2003)

The worldwide occurrence of MCs (Table 1.1) was demonstrated in various cyanobacterial genera, including the planktonic *Microcystis* sp. (e.g. *M. aeruginosa, M. wesenbergii, M. viridis*), *Planktothrix* sp. (e.g. *P. rubescens, P. agardhii*), *Anabaena* sp. (e.g. *A. flos-aquae*), benthic *Oscillatoria* sp. (e.g. *O. limosa*) and less frequently by *Nostoc* sp., *Anabaenopsis* sp. and *Hapalosiphon* sp. (Sivonen and Jones, 1999; van Apeldoorn et al., 2007). It is noteworthy to mention that the most widely distributed genera in the Northern Hemisphere, especially in late summer and autumn are colonial unicellular *Microcystis* preferring warmer waters as well as the filamentous *Planktothrix* (Mur et al., 1999). Indeed, *P. rubescens* is known to occur at high cell densities under ice and in clear lakes (Falconer, 2005a). However, "It is likely that the list of confirmed toxic species will increase in the future due to the isolation of new species and starins, and because of the use if improved isolation, culturing and analytical methods" (Sivonen and Jones, 1999).

Microcystis aeruginosa, a non-nitrogen fixing species, is most dominant in nutrient-rich environments, although it was also detected at high cell densities in less polluted waters (Sivonen and Jones, 1999). Additionally, benthic *Oscillatoria limosa* was identified e.g. in alpine lakes from Switzerland (Mez et al., 1996; Mez et al., 1997; Mez, 1997a). The latter observations clearly demonstrate that MC producing species are not only abundant in eutrophic waters but also in oligotrophic lakes. Nevertheless, high phosphorus and nitrogen levels trigger cyanobacterial strains to produce higher amounts of toxins (van Apeldoorn et al., 2007).

1. Cyanobacteria – a general introduction

Table 1.1: A summary of worldwide detected cyanobacterial species

Location	Cyanobacterial Species	Reference(s)
Finland	*Anabaena flos-aquae, A. circinalis, A. lemmermannii*	(Sivonen *et al.*, 1990)
Norway	*Anabaena flos-aquae*	(Sivonen *et al.*, 1992)
Denmark	*Microcystis botrys, Planktothrix agardhii, P. mougeotii*	(Henriksen, 1996)
England	*Nostoc sp.*	(Beattie *et al.*, 1998)
Switzerland	*Oscillatoria limosa*	(Mez *et al.*, 1996)
Germany	*Planktothrix rubscens*	(Ernst *et al.*, 2001)
Greece	*Anabaenopsis millerii*	(Lanaras *et al.*, 1989; Lanaras and Cook, 1994)
France	*Microcystis sp.*	(Sevrin-Reyssac and Pletikosic, 1990)
Japan	*Microcystis viridis, M. Aeruginosa, M. wesenbergii, Aphanizomenon flos-aquae*	(Kusumi *et al.*, 1987; Xie *et al.*, 2007)
Canada	*Anabaena flos-aquae*	(Krishnamurthy *et al.*, 1989; Kotak *et al.*, 1993)
Brazil	*Microcystis aeruginosa*	(Azevedo *et al.*, 1994)
China	*Micocystis sp., Anabaena sp*	(Chen *et al.*, 2005)
S. Africa	*Microcystis aeruginosa*	(Oberholster *et al.*, 2009)
Australia	*Microcystis sp., Oscillatoria sp., Pseudoanabaena sp., Aphanocapsa sp.*	(Kankaanpaa *et al.*, 2005)
USA (CA)	*Microcystis aeruginosa*	(DeVries *et al.*, 1993)

Indeed many cyanobactertial species are capable of producing several MCs at the same time, but usually only one or two congeners in any single strain (Sivonen and Jones, 1999). However, it seems to be the case that special MC congeners are typical for certain cyanobacterial taxa, like one major demethylated MC in populations of *Planktothrix rubescens* from German field samples (Fastner *et al.*, 1998). The most often occurring MC congener is suggested to be MC-LR, but likely simply due to the fact that MC-LR was the first commercially available chemical standard for analytical detection methods (Sivonen and Jones, 1999). MC-LR was reported by several authors to represent the most abundant variant in bloom and strain samples from Portugal (Vasconcelos *et al.*, 1996), France (Vezie *et al.*, 1998), Canada (Kotak *et al.*, 1993a) and Japan with co-occurring MC - RR and –YR produced by *Microcystis aeruginosa* (Watanabe *et al.*, 1988; Sivonen and Jones, 1999).

Information about MC concentrations in surface waters and scums has been reported only recently, due to new analytical methods like LC-MS, HPLC, ELISA and PP inhibition assay. The highest published MC concentration in blooming samples were 200 – 7300 µg MC-LR and MC-RR /g dry

1. Cyanobacteria – a general introduction

weight (dw) and 1000 – 7100 µg MCs /g dw, collected from China (Zhang et al., 1991) and Portugal (Vasconcelos et al., 1996), respectively (Sivonen and Jones, 1999). In water samples, highest concentrations were reported from Germany with 1 – 25000 µg MCs /l (WHO, 1998; Sivonen and Jones, 1999).

MCs are extremely stable and resistant to chemical hydrolysis or oxidation at near neutral pH and still remain potent after boiling. Rapid chemical hydrolysis could only be observed under laboratory conditions (Sivonen and Jones, 1999). Indeed, ozone or other strong oxidizing agents are useful methods to oxidize MCs whereas rapid degradation requires UV light close the absorption maxima of MCs (Tsuji et al., 1995; Sivonen and Jones, 1999; Hitzfeld et al., 2000a). However, under natural condition in the presence of pigments (e.g. phycobiliproteins) and full sunlight, MC breakdown (>90%) was reported to take two to six weeks, depending on the photopigment concentration (Sivonen and Jones, 1999). So far, biodegradation was only demonstrated by bacteria of the genus Sphingomonas and more recently Paucibacter toxinivorans and Sphingosinicella microcystinivorans (Park et al., 2001; Maruyama et al., 2003; Rapala et al., 2005a). Therefore, it is not surprising that MCs survive several weeks to months in sterile water without any degradation but in conjunction with natural waters (e.g. rivers, lakes) MCs breakdown can be observed already after 6 – 20 days, depending on the MC congener and the water source (Edwards et al., 2008).

The uptake of dissolved MCs (e.g. in drinking water) or ingested cyanobacterial cells containing MCs (e.g. recreational activities) is mainly dependent on their hydrophilic character and structural size. The latter facts make these toxins unable to cross cell membranes by passive diffusion (Eriksson et al. 1990; Fischer et al. 2005; Komatsu et al. 2007). Rather, they require a transport system to enter the venous blood (portal vein via the epithelium of the ileum) and subsequently cells (e.g. hepatocytes) (Eriksson et al., 1990; Runnegar et al., 1995). However, Eriksson et al. (Eriksson et al., 1987) demonstrated that a peptide toxin from the cyanobacterium *Microcystis aeruginosa* revealed a cell-type specific cytotoxicity. Three years later it could be revealed by the same authors that a multispecific transport system for bile acids is involved in the uptake of MC-LR into primary rat hepatocytes (Eriksson et al., 1990). Runnegar et al. (Runnegar et al., 1995) suggested Na^+ independent transporters, called organic anion transporting polypepetides (according to the new nomenclature (Hagenbuch and Meier, 2004); human OATP, rodent Oatp and murine mOatp) to be involved in the sinusoidal uptake of MCs.

Only in the last years cloning experiments and over expressing cell systems have clearly indicated OATPs/Oatps capable for transporting MCs. However, only MC-LR uptake was indicated so far by over expressing *Xenopus leavis* oocytes (OATP1A2, 1B1, 1B3 and skate Oatp1d1) with Km values of 20 ± 8 µM, 7 ± 3 µM, 9 ± 3 µM and ~27 µM, respectively (Fischer et al., 2005; Meier-Abt et al.,

2007). Rat Oatp1b2 expressing *Xenopus leavis* oocytes accumulated MC-LR about 2 fold compared to controls, but no uptake could be observed for rat Oatp1a1, Oatp1a4 and OATP2B1 (Fischer *et al.*, 2005). In addition, uptake of MC-LR was recently confirmed via stably transfected HEK293 cell line (OATP1B1 and 1B3) with similar cytotoxicity for both overexpressed transporters (IC_{50} = 6.6 and 6.5 nM, respectively) (Komatsu *et al.*, 2007). A Km value of 1.2 µM MC-LR was only investigated for HEK293-OATP1B3 (Komatsu *et al.*, 2007). Transiently transfected HeLa cells (OATP1B1 and 1B3) are over 1000 fold more sensitive to MC-LR than the control transfected cells (Monks *et al.*, 2007). In addition, single MC congeners (MC-LR, -LW, -LF, -YR and –RR) showed variable cytotoxicities in OATP1B1 and 1B3 transfected cells, including MC-LF and MC-LW with IC50 < 1nmol/l (Monks *et al.*, 2007). More recently, an *in vivo* study using mOatp1b2 knockout mice revealed lack of hepatotoxicty in MC-LR exposed animals compared with wild type mice (Lu *et al.*, 2008). Interestingly, MC-LR immunoblotting revealed a positive band corresponding to the catalytic subunits of PP1 and PP2A for both MC-LR treated wild-type and mOatp1b2$^{-/-}$ mice, although the signal was less intense for the knockout mouse (Lu *et al.*, 2008).

The mode of action of MCs *in vitro* and *in vivo* leading to the observed toxicity is mainly due to the potent inhibition of ser/thr-specific PPs, mainly PP1 and PP2A (Honkanen *et al.*, 1990; MacKintosh *et al.*, 1990; Hastie *et al.*, 2005) but is also true for PP4 and PP5 (Hastie *et al.*, 2005). PP3, formerly called PP2B is only weakly inhibited (MacKintosh, 1993). The formation of covalent MC-PP adducts represents a separate event from the inhibition of PP activity by this toxin (Runnegar *et al.*, 1995a).

Determination of the crystal structure of MC-PP1α adducts revealed new insights on the molecular level (Goldberg *et al.*, 1995). Binding of MCs to the catalytic subunit of PPs is mediated by three events (MacKintosh *et al.*, 1995; Gulledge *et al.*, 2003; Hastie *et al.*, 2005; Maynes *et al.*, 2006). First of all, the long hydrophobic Adda chain (position 5) is placed in the hydrophobic groove of PP1α, close to the active side. Next and most important for cyclic peptide toxicity (including nodularins) is the indirect interaction of glutamyl carboxylate (position 6) via sharing two water molecules with two catalytic metal atoms of the PP. In addition, the carboxyl group of MeAsp residue (position 3) of MC-LR interacts with Arg96 and Tyr134 of PP1α (Maynes *et al.*, 2006). Finally, the double bond of the Mdha residue (position 7) interacts with Cys273 (PP1α) and/or Cys266 (PP2Aα) to form a covalent linkage (MacKintosh *et al.*, 1995; Runnegar *et al.*, 1995a). While the initial non-covalent binding takes minutes, subsequent covalent and therefore irreversibly binding of MCs to the catalytic subunit of PPs requires several hours, without further strengthening of PP inhibition (Goldberg *et al.*, 1995; Craig *et al.*, 1996). The lack of Mdha in the structure of nodularin resulted in a non-covalent binding and subsequent PP inhibition (Bagu *et al.*, 1997) but with a similar LD_{50} values compared to e.g. MC-LR (Sivonen and Jones, 1999; van Apeldoorn *et*

1. Cyanobacteria – a general introduction

al., 2007). The Adda-glutamate region (position 5 and 6) is well conserved within the group of MCs and nodularins and therefore similar if not equal LD_{50} were indentified with values between 50 µg/kg (e.g. nodularin, MC-LR and also true for MC-LA (Maynes *et al.*, 2006) lacking covalent interaction with Cys273 of PP1a) to 90 µg/kg (e.g. MC-LY) (Sivonen and Jones, 1999). However, the latter observation can not explain variations in LD_{50} values of 250 µg/kg for MC-AR up to 600 µg/kg for MC-RR compared with MC-LR (Sivonen and Jones, 1999).
In addition to the main target of MC, another mode of action was suggested by Mikhailov *et al* (Mikhailov *et al.*, 2003) indicating MCs capable for binding to ATP synthetase, which furthermore could be associated with the apoptotic potential of MCs. MCs were also demonstrated to inhibit the protein synthesis which proceeds with the activation of glucose-6-phosphatase (Claeyssens *et al.*, 1993; Claeyssens *et al.*, 1995).

PPs are ubiquitously distributed enzymes, involved in fundamental cell signaling processes regulating many crucial biological functions, including neurotransmission, learning and memory, and cell division (McCluskey *et al.*, 2001; McCluskey *et al.*, 2002). Thus, and not surprising, MC intoxicated species, including humans indicate severe pathology.
On the molecular level, one major consequence of MC induced PP inhibition (*in vitro* and *in vivo*) is mainly due to the disruption of the cellular phosphorylation/dephosphorylation balance, targeting cytosolic, cytoskeletal and cytoskeletal-associated proteins (Hooser *et al.*, 1989; Hooser *et al.*, 1991; Ghosh *et al.*, 1995; Toivola and Eriksson, 1999; Komatsu *et al.*, 2007). At moderate MC concentrations, the first visible effect is a dramatic reorganization of both the intermediate filaments (due to hyperphosphorylation of keratin 8 and 18) and microfilaments, resulting in a loss of cell morphology, cell adhesion and finally cellular necrosis (Eriksson *et al.*, 1989; Hooser *et al.*, 1989; Falconer and Yeung, 1992; Ohta *et al.*, 1992; Toivola *et al.*, 1997; Toivola *et al.*, 1998; Toivola and Eriksson, 1999). The latter observation was also confirmed for degenerating actin filaments in MC-LR exposed primary human hepatocytes (Batista *et al.*, 2003) and rat hepatocytes *in vitro* and *in vivo* (Hooser *et al.*, 1991). More recently, kinetics of cytoskeletal reorganization in various cell-types was demonstrated and revealed initially a breakdown of intermediate filaments, followed by a rapid disorganization of microtubules, with an later onset of microfilament alterations (Wickstrom *et al.*, 1995; Khan *et al.*, 1996).
Next to cytoskeleton degeneration after MC exposure, two events, namely free radical formation and mitochondria alterations were observed *in vitro* (including primary hepatocytes, lymphocytes and in a wide variety of cell lines) and *in vivo*, thus triggering cytotoxicity and the induction of apoptotic pathways (Ito *et al.*, 1997; Ito *et al.*, 1997a; Mankiewicz *et al.*, 2001; Ding and Nam Ong, 2003; Lankoff *et al.*, 2003; Botha *et al.*, 2004; Gehringer, 2004; Lankoff *et al.*, 2004a; Fu *et al.*,

2005; Teneva *et al.*, 2005; Weng *et al.*, 2007). Indeed, apoptosis is an essential physiological process for the development and maintenance of multicellular organisms (Leist and Jaattela, 2001). Many eukaryotic cells that die undergo a programmed cell death including the above described morphological alterations but also biochemical changes, e.g. via the activation of special proteases, namely caspases (Leist and Jaattela, 2001). However, detailed information about MC triggered death pathways need to be further investigated (Gehringer, 2004; Xing *et al.*, 2008), especially in terms of caspase dependent apoptosis with so far contradictory reports (Fladmark *et al.*, 1999; Ding *et al.*, 2002; Gehringer, 2004; Komatsu *et al.*, 2007). Altered protein expression of p53, Bcl-2 and Bax were recently reported after MC-LR exposure, in human amniotic cells FL and in primary rat hepatocytes (Fu *et al.*, 2005; Xing *et al.*, 2008). Both *in vitro* and *in vivo* studies demonstrated an increase of p53 and Bax proteins and a decrease of Bcl-2, only observed *in vitro* (Fu *et al.*, 2005; Xing *et al.*, 2008). Additionally, it was reported that MC-LR can induce p53 hyperphosphorylation in vitro, which in turn triggers p53-dependent apoptosis (Guzman *et al.*, 2003; Fu *et al.*, 2005). More recently, two *in vitro* studies have observed after MC-LR induced PP inhibition, activation of the Ca^{2+}/Calmodulin-dependent protein kinase II, executing various parameters of cell death (Fladmark *et al.*, 2002; Krakstad *et al.*, 2006). Finally, Komatsu *et al* (Komatsu *et al.*, 2007) have reported that OATP1B1 and 1B3 transfected HEK293 cells treated with MC-LR revealed phosphorylation of mitogen activated protein kinases (MAPKs) such as extracellular signal-regulated protein kinases (ERK1/2), c-Jun NH2-terminal kinases (JNK) and stress activated protein kinases (p38). Proinflammatory cytokines and ROS are stimuli derived by MC treatment (Gehringer, 2004) thus leading to the activation not only of p38 but also of JNK. The natural membrane permeable PP inhibitor okadaic acid was reported to activate ERK1/2, JNK and p38 (Rossini *et al.*, 1999; Boudreau and Hoskin, 2005; Yoon *et al.*, 2006; Boudreau *et al.*, 2007).

In laboratory studies, acute toxicity was observed in mice and rats with oral LD_{50} values of 5000 µg MC-LR /kg, >5000 µg MC-LR /kg, respectively (Fawell *et al.*, 1999). Clinical signs included hypoactivity and piloerection with significant microscopic changes including dark discoloration of the liver with diffuse hemorrhage, marked centrilobular hemorrhage, centrilobular necrosis and cytoplasmic vacuolation (Fawell *et al.*, 1999). Differences in acute toxicity could be demonstrated between oral exposure and i.p. injections varying between 30 to 100-fold (Fawell *et al.*, 1999). Interestingly, i.p. treated mice demonstrated variability in LD_{50} values from 25 µg MC-LR /kg bw to 150 µg/kg bw (Meriluoto *et al.*, 1990; Miura *et al.*, 1991; Fawell *et al.*, 1999; van Apeldoorn *et al.*, 2007), whereas a value of 50 µg MC-LR /kg bw is commonly accepted (Sivonen and Jones, 1999; van Apeldoorn *et al.*, 2007). MC-LR treated mice (i.p.) exhibited clinical signs of prostration, convulsions and slow respiration with severe liver damage characterized by disruption of liver cells-

and sinusoidal structure and increased liver weight due to hepatic hemorrhage (Fawell et al., 1999). Next to hepatotoxicity, dark discoloration of the kidney, spleen, glands and thymus was additionally observed (Fawell et al., 1999). As already mentioned, LD_{50} values differ depending on the application route, employed species and strains but especially among different MC congeners (e.g. MC-LR vs. MC-RR) (Sivonen and Jones, 1999; Zurawell et al., 2005).

In a subchronic/chronic in vivo study, MC-LR was administered orally to male and female mice at 0, 40, 200 and 1000 µg/kg bw per day for 2 and 13 weeks (Fawell et al., 1994; Fawell et al., 1999). However, in the 2 week study, the only findings were microscopic changes in the liver in four male and two female mice at 1000 µg/kg bw (Fawell et al., 1994; van Apeldoorn et al., 2007). In the 13 week gavage study the NOEL was 40 µg/kg bw per day, based on pathological changes in the liver (Fawell et al., 1999). At the highest dose (1000 µg/kg bw) all mice revealed liver damage, including chronic inflammation. Similar to the study of Fawell et al (Fawell et al., 1999), Heinze (Heinze, 1999) reported dose-dependent increase in liver weights, elevated serum activities of e.g. of lactate dehydrogenase and liver pathology in mice treated with 0, 50 and 150 µg MC-LR /kg bw over 4 weeks via drinking water. An additional chronic study using Balb/c mice was performed, mimicking a likely human exposure scenario (Ueno et al., 1999). Animals were treated over 18 months with 0 and 20 µg MC-LR /l of drinking water. The mean cumulative MC-LR intake after 18 months was estimated at 35.5 µg per mouse. No chronic toxicity and no accumulation of MC-LR in the liver were observed (Ueno et al., 1999).

1.4. Microcystin poisonings – a risk to public health

The environmental and health consequences of cyanobacterial booms for wild and domestic animals as well as for humans, mainly depend on the ability of extreme proliferating cyanobacteria to produce toxic secondary metabolites. Chances of MC poisoning increase with bloom density, towards the end of a season (e.g. senescence during late summer) and after lysis of cyanobacterial cells (Dittmann and Wiegand, 2006). Public and scientific concerns resulted in huge amounts of scientific publications and media reports which led to a special issue of Toxicology and Applied Pharmacology 203 (2005) with the Special Issue Introduction "Risk cyanobacterial toxins: Occurrence, ecology, detection, toxicology, and health effects assessment".

However, numerous cases of animal poisonings have been reported associated with MCs (Briand et al., 2003; Stewart et al., 2008), including morbidities and mortalities in rhinoceros (e.g. Kruger National Park) (Oberholster et al., 2009), wild birds and turtles (Matsunaga et al., 1999; Park et al., 2001a; Krienitz et al., 2002), cattle (Naegeli et al., 1997), fish and crustaceans (Magalhaes et al., 2001; Magalhaes et al., 2003; Cazenave et al., 2005), sheep (Carbis et al., 1994; Carbis et al., 1995)

1. Cyanobacteria – a general introduction

and dogs (DeVries *et al.*, 1993). Animal fatalities are not the scope of this chapter, rather pointing out likely human exposure scenarios, consequent health effects and case reports of human intoxications.

Human MC intoxications occur most likely via oral- and/or by inhalation exposure either voluntarily or accidentally e.g. by:

- Contaminated drinking water (Teixera *et al.*, 1993; WHO, 1999; Hitzfeld *et al.*, 2000a; Falconer and Humpage, 2005; Falconer, 2005a).
- Recreational activities in blooming waters (e.g. swimming, canoeing) (Dietrich *et al.*, 2008; Backer *et al.*, 2008; Pilotto *et al.*; 1997).
- Contaminated food (e.g. fish, prawns) (Ernst *et al.*, 2000; Magalhaes *et al.*, 2003; Kankaanpaa *et al.*, 2005; Ibelings and Chorus, 2007).
- Contaminated blue-green algae food supplements (BGAS) (Schaeffer *et al.*, 1999; Gilroy *et al.*, 2000; Lawrence *et al.*, 2001a).

This is especially true for short-terms effects in humans (Table 1.2) if drinking water reservoirs show high densities of toxic cyanobacteria species (e.g. *Microcystis* sp. and *Anabaena* sp.) (Teixera *et al.*, 1993 1998; Hitzfeld *et al.*, 2000a; Carmichael *et al.*, 2001; Fleming *et al.*, 2002; Hoeger *et al.*, 2004; Falconer and Humpage, 2005; Falconer, 2005a 1999) associated with either a natural bloom breakdown or by artificial cell lysis (e.g. copper sulphate), thus resulting in the release of MCs (Kuiper-Goodman *et al.*, 1999).

The most fatal case in conjunction with MCs in drinking water occurred in the Paulo Afonso region of Bahia State, Brazil when a newly flooded dam (Itaparica Dam reservoir, 1988) demonstrated an intense bloom. In this bloom, *Anabaena* sp. and *Microcystis* sp. genera were present in untreated water with cell densities of approximately 100000 – 1000000 cells/ml. Within 42 days, more than 2000 cases of gastroenteritis were reported (>70% children under 5 years of age) only associated with people drinking boiled water from the Itaparica Dam reservoir, leading to death of 88 persons that included mostly children (Teixera *et al.*, 1993).

Inhalation of cyanobacterial toxins and consequent acute or chronic exposure could be due to habitation near routinely contaminated surface waters, use of contaminated water for hygienic purposes and finally during recreational activities (Dietrich *et al.*, 2008), reported from Finland (Rapala *et al.*, 2005), Canada (Dillenberg and Dehnel, 1960), England (Turner *et al.*, 1990) and in the US, especially Florida (Burns, 2004; Backer *et al.*, 2008).

1. Cyanobacteria – a general introduction

Table 1.2: Summary of acute MC poisonings – human case reports

Location	Year	Application	Symptoms	Source	Ref
Ohio, USA	1931	drink. w.	5000-8000 cases of gastroenteritis	?	1
Saskatchewan, Canada	1959	w. activities	13 people with painful diarrhoea, nausea, muscular pains	Microcystis, Anaba.	2
Harare, Zimbabwe	1960-65	drink. w.	gastroenteritis	Microcystis	3
Sewickley, USA	1976	drink. w.	gastroenteritis, 62% of the population	?	4
Armidale, Australia	1983	drink. w.	liver damage	Microcystis	5
Itaparica, Brazil	1988	drink. w.	>2000 cases of gastoenteritis, 88 deaths	Anaba., Microcystis	6
Staffordshire, England	1989	w. activities	10 army recruits show abdominal pain, pneumonia, dry cough	M. aeruginosa, E. coli	7
Australia	1995	w. activities	852 people with vomiting, diarrhoea, mouth ulcers, flu symptoms	Cyanobacteria	8
Caruaru, Brazil	1996	dialysis i.v.	131 patients developed neurotoxicity, liver failure, 76 deaths	MC-LR, -YR, -RR	9
Scania, Sweden	?	drink. w.	121 cases of gastroenteritis	P. agardhii	10
Finland	?	hot steam (sauna)	48 cases of gastroenteritis, dermal and neurological symptoms	Cyanobacteria	11
Florida, USA	?	hot steam (shower)	?	MC, anatoxin-a, cylindrospermopsin	12

1: (Tisdale, 1931); 2: (Dillenberg and Dehnel, 1960); 3: (Zilberg, 1966); 4: (Lippy and Erb, 1976); 5: (Falconer et al., 1983; Falconer, 2005a); 6: (Teixera et al., 1993); 7: (Turner et al., 1990); 8: (Pilotto et al., 1997); 9. (Pouria et al., 1998; Carmichael et al., 2001; Azevedo et al., 2002); 10: (Annadotta et al., 2001); 11: (Hoppu, 2002); 12: (Burns, 2004); water (w.); Anabaena (Anaba.)

Chronic exposure to drinking water contaminated with low MC concentrations represents the most likely and "uncalculable" risk scenario for humans. However, toxin distribution, accumulation, biotransformation and elimination have to be considered for an adequate human risk assessment. A systematic screening of drinking water sources (e.g. pond, ditch water) in China, suggested cyanobacterial toxins together with hepatitis B and aflatoxin B1 from food items, may likely contribute to the high incidences of primary liver and colon cancer in this region (Harada et al., 1996; Ueno et al., 1996; Harada and Ueno, 1996a; Kuiper-Goodman et al., 1999).

1. Cyanobacteria – a general introduction

Volume 94 of the International Agency for Research on Cancer (IARC) monographs, classified MC-LR as possible carcinogenic to humans (group 2B) (Grosse et al., 2006) mainly based on the below described tumor promoting effects in laboratory animals (Falconer, 2008).

This was confirmed by recent *in vitro* observations which demonstrated expression of selected genes after MC-LR exposure, known to be involved in the cell response to DNA damage and apoptosis (Zegura et al., 2008). However, MCs were demonstrated to induce reactive oxygen species (ROS) and lipid peroxidation with subsequent and therefore secondary oxidative DNA damage (Zegura et al., 2003; Bouaicha and Maatouk, 2004; Sano et al., 2004; Zegura et al., 2004). In a recent *in vivo* study it could also be demonstrated that ROS play a critical role in MC-LR induced hepatocyte apoptosis (Weng et al., 2007) which was suggested to cause the observed DNA damage due to cytotoxicity rather than genotoxicity (Lankoff et al., 2004).

Some evidence derived from *in vitro* and animal studies suggest MCs as tumor promoting toxins (Dietrich and Hoeger, 2005; Dittmann and Wiegand, 2006; Humpage, 2008). Cultured hepatocytes appear to suppress apoptosis and to trigger cell proliferation at low MC concentrations (Humpage and Falconer, 1999) resulting in tumor promotion (Ohta et al., 1992). It was suggested by Humpage (Humpage, 2008) that these effects may be linked to the enhancement of the growth of hepatic and colonic pre-cancerous lesions in animal models (Fujiki, 1992; Nishiwaki-Matsushima et al., 1992; Fujiki and Suganuma, 1993; Fujiki et al., 1996; Ito et al., 1997; Ito et al., 1997a; Humpage et al., 2000). Indeed, as already mentioned above, chronic exposure of humans have more recently been linked to liver and colon cancer incidences in China (Yu, 1989; Yu, 1994; Yu, 1995; Zhou et al., 2000; Zhou et al., 2002) and an elevated risk was found for residents near surface water treatment plants in Florida (Fleming et al., 2002).

As many human fatalities were reported in conjunction with MC contaminated drinking water, the WHO derived in 1998 a provisional guideline value for only MC-LR in drinking water of 1 µg/l for 60 kg person consuming 2 l of water per day (WHO, 1999). The WHO concluded that there are insufficient data to derive a guideline for other MC congeners, one major problem of MC risk assessment. A detailed current review of MCs in drinking water and their associated problems regarding the derived guidance value can be found in Dietrich and Hoeger (Dietrich and Hoeger, 2005), Falconer and Humpage (Falconer and Humpage, 2005) and more recently in Dietrich et al (Dietrich et al., 2008).

1.5. Organic anion transporting polypeptides

Transporters of the solute carrier (SLC) superfamily represent integral membrane proteins, acting as passive transporters, exchanger of nutrients (e.g. amino acids, sugars) and ion-coupled transporters, involved in the uptake of inorganic ions, organic ions, including numerous drugs (Nies, 2007).

There is a great pharmacological interest for SLC transporters as well as for members of the ATP-binding cassette (ABC) superfamily of transporters in terms of drug (e.g. chemotherapeutics like methotrexate and cytarabine) uptake and efflux, respectively (Nies, 2007). Because many pharmacological substances poorly cross barriers (e.g. the BBB/BCSFB) and membranes (e.g. tumor cells) a detailed understanding of potent drug transporters seems to be crucial in conjunction with chemotherapy resistance.

In the human genome 360 SLC carriers have been found, which are grouped into 46 families (Nies, 2007).

Some family members of the solute carrier:

- H^+/oligopeptide cotransporter — *SLCO15A1 – SLC15A4*
- The organic cation transporters (OCTs) — *SLC22A2 and SLC22A3*
- Na^+-coupled nucleoside transporter — *SLC28A2*
- Reduced folate transporter — *SLCO19A1*
- Organic anion transporting polypeptide (SLCO) — *SLCO/SLC21 family*

Organic anion transporting polypeptides (human OATP/*SLCO* / rodent Oatp/*Slco*; OATP/Oatp represent the corresponding protein and *SLCO/Slco* represent the corresponding gene) are a grown gene superfamiliy of polyspecific sodium independent solute membrane carriers mediating the uptake of a broad spectrum of amphipathic organic substrates such as bile salts (e.g. taurocholate, glycocholate, cholate), hormones and their conjugates (e.g. DHEAS, cortisol, T_3, T_4, estrone sulfate), drugs (e.g. methotrexate, pravastatin, digoxin), antibiotics (e.g. rifampicin), other organic anions (e.g. folate, BSP) and natural toxins like phalloidin, and ochratoxin A, including MCs (Hagenbuch and Meier, 2003; Hagenbuch and Meier, 2004; Fischer *et al.*, 2005; Konig *et al.*, 2006; Komatsu *et al.*, 2007; Monks *et al.*, 2007). Currently 80 carriers of the OATP/Oatp superfamiliy in at least 13 different species have been identified with 36 representing members in human, mouse and rat (Hagenbuch, 2007). Mammalian OATPs/Oatps are expressed in multiple organs like the liver, kidney, intestine, testis, placenta, BBB and blood-cerebrospinal fluid barrier (BCSFB) with either a ubiquitous tissue distribution (e.g. OATP1A2 located in the liver, kidney and brain) or an exclusively expression, for instance for OATP1B1 and 1B3 in the liver, OATP4C1 in the kidney

1. Cyanobacteria – a general introduction

and OATP6A1 in the testis (Hagenbuch and Meier, 2004). In the brain, several OATPs/Oatps are expressed at BBB and at BCSFB involved in both uptake and efflux of numerous substances (Westholm et al., 2008; Nies, 2007), including human OATP1A2, 2B1, 3A1 and most likely OATP4A1 and 1C1 (Table 1.3).

In addition to human OATPs located in the brain, so far five rodent Oatps, namely OATP1a1, 1a4, 1a5, 1c1, 2a1 have been identified at the protein level at the BBB/BCSFB (see Table 3) and is possibly also true for the ubiquitously expressed Oatp3a1 and 4a1 (Westholm et al., 2008).

Table 1.3: OATPs/Oatps at the BBB/BCSFB

Protein	Localisation	Reference(s)
Oatp1a1	protein; CP, am	1
Oatp1a4	protein; CP: bm; BBB: lum, abm	2
Oatp1a5	mRNA; CP: am and protein; likely	3, 4
Oatp1c1	protein; CP: bm; BBB: lum, ab	5
Oatp2a1	protein; CP and BBB: predominantly lum	6
OATP1A2	protein; BBB: lum, abm ?; CP: ?	7, 8, 9
OATP2B1	protein; BBB: lum	10
OATP3A1	protein; CP: bm and membranes of glia cells and neurons	11
OATP4A1	brain but not in BBB	12, 10
OATP1C1	several regions of the brain	13

CP: choroid plexus epithelial cells; BBB: endothelial cells of the blood-brain-barrier; am: apical membrane; bm: basolateral membrane; lum: luminal membrane; abm: abluminal membrane; 1: (Angeletti et al., 1997); 2: (Gao et al., 1999); 3: (Ohtsuki et al., 2004); 4: (Hagenbuch and Meier, 2004); 5: (Sugiyama et al., 2003); 6: (Kis et al., 2006); 7: (Gao et al., 2000); 8: (Lee et al., 2005); 9: (Nies, 2007); 10: (Bronger et al., 2005); 11: (Huber et al., 2007); 12: (Sato et al., 2003); 13: (Pizzagalli et al., 2002).

"Polygenetic classification and new nomenclature of human and rodent (rat, mouse) members of the OATP/*SLCO* superfamiliy of membrane transporters. Oatps/OATPs with amino acid sequence identities ≥40% among each other belong to the same OATP family (e.g. OATP1, OATP2, OATP3, OATP4, OATP5, OATP6). Proteins with amino acid sequence identities ≥60% are grouped into subfamilies and denoted with *capital letter* after the family number (e.g. OATP1A, OATP1B, OATP1C, OATP2B, etc.). Individual proteins (*genes*) are continuously numbered according to the chronology of their identification. The "Oatp" (rodents) / "OATP" (human) symbols denote proteins, the "*Slco*" / "*SLCO*" symbols indicate the respective *genes*. Mouse Oatps are indicated by "m". The tree was calculated using the ClustalW program (http://www.ebi.ac.uk/clustalw/) and

1. Cyanobacteria – a general introduction

visualized using the program TREVIEW (Page, 1996). Only the families OATP1 to OATP6 are indicated, since only these six families contain human OATPs" (Hagenbuch and Meier, 2004).

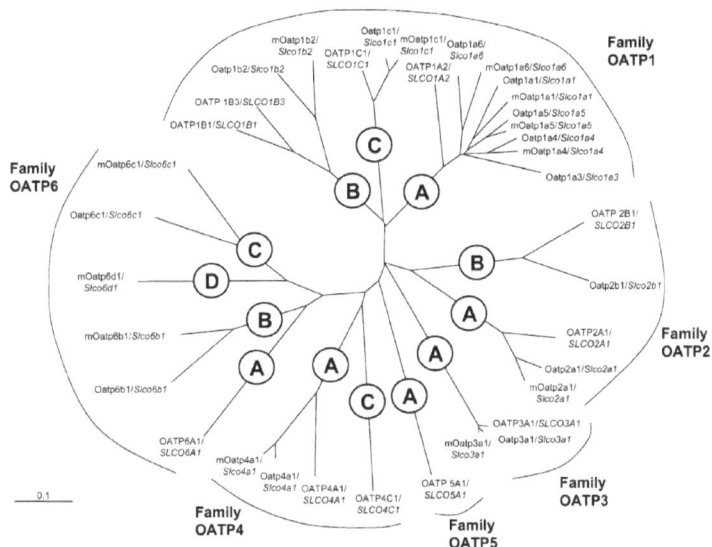

Figure 1.6: Polygenetic tree of OATPs/Oatps
(Hagenbuch and Meier, 2004)

Drug uptake, elimination and selected tissue penetration are widely recognized to be mainly dependent on expressed and functional uptake and efflux transporters (Nies, 2007). Next to the extensive knowledge of drug efflux transporters (e.g. MDR1) only recent studies have implicated the importance of drug uptake transporters (e.g. OATPs/Oatps) into closer considerations, especially important to understand the overall drug disposition process. Therefore, cell-type specific expression of OATPs/Oatps in organs like the liver, intestine, kidney and brain, seems to alter the extent of drug bioavailability (Lee et al., 2005).

More recently it could be demonstrated that genetic variations (single nucleotide polymorphisms) in e.g. *SLCO1A2*/OATP1A2 among different ethnicities and/or individuals may have important implications to the disposition and tissue penetration of drugs (Lee et al., 2005). It was finally concluded by Lee et al. (Lee et al., 2005) that the observed polymorphisms in *SLCO1A2*/OATP1A2 may be an important and yet unrecognized contributor to inter-individual variability in drug disposition and central nervous system uptake of substrate drugs.

In addition, splice variants of e.g. OATP3A1 (OATP3A1_v1 and OATP3A1_v2) isolated from the human brain demonstrated varying expression patterns (Huber et al., 2007). For instance, OATP3A1_v1 (710 amino acids) was presented to be ubiquitously expressed, including membranes of neuroglia, whereas OATP3A1_v2 (692 amino acids) exhibited predominant expression in Sertoli cells in the testis and in cell bodies and axons of neurons (Huber et al., 2007).

1.6. Initial evidence for the neurotoxic potential of microcystins – history and goal of the study

The fact that MCs are called "hepatotoxins" by many scientists in the field is based on the organ distribution of MCs, primarily, if not exclusively targeting the liver (Dietrich and Hoeger, 2005). Indeed, most acute MC intoxications of humans and animals demonstrate liver pathology (Hooser et al., 1989; Pouria et al., 1998; Fawell et al., 1999; Beasley et al., 2000; Hooser, 2000; Azevedo et al., 2002). This is most likely due to the hepatic "first-pass-effect" and additionally high level of functionally expressed OATPs/Oatps in hepatocytes capable for transporting MCs (Eriksson et al., 1990; Fischer et al., 2005; Komatsu et al., 2007; Monks et al., 2007).

Moreover, OATPs/Oatps are not only expressed in the liver, rather in all tissues/organs, including the gastrointestinal tract, kidney, BBB and BCSFB (Hagenbuch and Meier, 2004; Nies, 2007; Westholm et al., 2008). Therefore, organ distribution of MCs is mainly governed by the presence or absence, type, expression and function of OATPs/Oatps capable for transporting single MCs (Dietrich and Hoeger, 2005).

Consequently, MCs can be detected in several tissues/organs, including the gastrointestinal tract (Botha et al., 2004a), the kidney (Nobre et al., 1999; Nobre et al., 2001; Milutinovic et al., 2002; Milutinovic et al., 2003; Nobre et al., 2004) as well as in the brain (Hall et al., 2008; Meriluoto et al., 1990; Nishiwaki et al., 1994).

Initial evidence for the neurotoxic potential of MCs was demonstrated more recently (Table 1.4), based on organ distribution studies (Hall et al., 2008; Meriluoto et al., 1990; Nishiwaki et al., 1994) and pathological observations in rodents (Falconer et al., 1988) as well as on neurological symptoms reported from intoxicated humans (Pouria et al., 1998; Carmichael et al., 2001; Azevedo et al., 2002) and rats (Maidana et al., 2006) that are likely associated with the presence of OATPs/Oatps at the BBB capable of transporting MCs into the brain (Fischer et al., 2005).

1. Cyanobacteria – a general introduction

Table 1.4: Summary of all published cases suggesting a neurotoxic potential of MCs

Organism (Strain)	Toxin	Exposure	Effect	Ref.
Male mice (NMRI)	^{3}H-dh-MC-LR	i.v. 24 h	2% brain, 70% liver, 6% kidney	1, 2
Female mice (ICR)	^{3}H-dh-MC-LR	p.o. 6 h	0.02% brain, 0.65% liver, 0.05% kidney	3, 2
Pregnant mice (CD-1)	^{3}H-dh-MC-LR	p.o. 4 h p.o. 48 h	1.06 ngMC/g brain, 0.71 ngMC/g liver, 1.09 ngMC/g kidney 0.55 ngMC/g brain, liver (nd), 0.57 ngMC/g kidney	4
Pregnant mice (Swiss Alb.)	*M. aeruginosa* extract	p.o.	10% of offsprings showed reduced brain size with an extensive damage in the outer region of the hippocampus (only one pup investigated)	5
Wister rats	*Microcystis* RST 9501 extract (D-Leu-MCLR)	i.hippoc. 15 min	ROS formation; short- and long-term memory alterations	6
Human	MC-LR, -YR, -AR	i.v.	89% of patients developed acute neurotoxicity (e.g. visual disturbance, deafness, tinnitus)	7, 8, 9
Human / *in vitro*	MC-LR		Human OATP1A2 located at the BBB mediate uptake of MC-LR	10
Fish (*Cyprinus carpio*)	*M. aeruginosa* extract	p.o. 72 h	Detection of MC-LR-protein phosphatase aducts in the brain	11

[%] denotes % of administered dose; i.v.: intravenous; p.o.: per oral; i.hippoc.: intrahippocampal; nd: not detected; 1: (Meriluoto et al., 1990); 2: (Dietrich and Hoeger, 2005); 3: (Nishiwaki et al., 1994); 4: (Hall et al., 2008); 5: (Falconer et al., 1988); 6: (Maidana et al., 2006); 7: (Pouria et al., 1998); 8: (Carmichael et al., 2001); 9: (Azevedo et al., 2002); 10: (Fischer et al., 2005); 11: (Fischer and Dietrich, 2000).

However, all the above described *in vivo* observations, lack direct evidence for a MC induced neuron-specific toxicity and additionally fail to take other MC congeners into account.

Therefore, the aims of this study were

- to assess in preliminary *in vitro* experiments evidence for single MC congener dependent neurotoxicity.
- to demonstrate more refined assessment of *in vitro* neurotoxicity in terms of PP inhibition, neurite degeneration and cytotoxicity using individual MC-LR, -LW and –LF.
- to investigate the apoptotic potential of single MC congeners.
- to develop an accurate and subtle assay for the determination of initial neurodegeneration induced by individual MC congeners at low concentrations.
- to characterize the presence of mOatps in membranes of primary murine neurons.
- to investigate the mOatp dependent uptake capabilities for individual MC congeners.
- to confirm the observed *in vitro* effects in a first *in vivo* experiment with mice.

1. Cyanobacteria – a general introduction

The overall approach of this work was to assess whether or not MCs represent potent neurotoxins *in vitro* and *in vivo* with a focus on neuron-specific toxicity induced by three different MC congeners.

2. Manuscript I

OATP-ASSOCIATED UPTAKE AND TOXICITY OF MICROCYSTINS IN PRIMARY MURINE WHOLE BRAIN CELLS

Feurstein D., Holst K., Fischer A., Dietrich D.R.

Human and Environmental Toxicology, University of Konstanz, Konstanz, Germany

Published in Toxicology and Applied Pharmacology 234 (2009) 247-255

Abstract

Microcystins (MCs) are naturally occurring cyclic heptapeptides that exhibit hepato-, nephro- and possibly neurotoxic effects in mammals. Organic anion transporting polypeptides (rodent Oatp / human OATP) appear to be specifically required for active uptake of MCs into hepatocytes and kidney epithelial cells. Based on symptoms of neurotoxicity in MC-intoxicated patients and the presence of Oatp/OATP at the blood-brain-barrier (BBB) and blood-cerebrospinal-fluid-barrier (BCFB) it is hypothesized that MCs can be transported across the BBB/BCFB in an Oatp/OATP-dependent manner and can induce toxicity in brain cells via inhibition of protein phosphatase (PP). To test these hypotheses, the presence of murine Oatp (mOatp) in primary murine whole brain cells (mWBC) was investigated at the mRNA and protein level. MC transport was tested by exposing mWBCs to three different MC-congeners (MC-LR, -LW, -LF) with/without co-incubation with the OATP/Oatp-substrates taurocholate (TC) and bromosulfophthalein (BSP). Uptake of MCs and cytotoxicity was demonstrated via MC-Western blot analysis, immunocytochemistry, cell viability and PP inhibition assays. All MC congeners bound covalently and inhibited mWBC PP. MC-LF was the most cytotoxic congener followed by -LW and -LR. The lowest toxin concentration significantly reducing mWBC viability after 48 h exposure was 400 nM (MC-LF). Uptake of MCs into mWBCs was inhibited via co-incubation with excess TC (50 and 500 µM) and BSP (50 µM). MC-Western blot analysis demonstrated a concentration-dependent accumulation of MCs. In conclusion, the *in vitro* data support the assumed MC-congener-dependent uptake in a mOatp-associated manner and cytotoxicity of MCs in primary murine whole brain cells.

Keywords: cyanobacteria, toxin, microcystin, protein phosphatase, Oatp/OATP, neurotoxicity, primary whole brain cells

2. Manuscript I

Introduction

Contamination of natural waters by cyanobacterial blooms represents a worldwide problem, causing serious water pollution and health hazards to humans and livestock. Human health problems are most likely associated with chronic exposure to low microcystin (MC) concentrations in poorly treated drinking water, contaminated food, e.g. fish, water snails, prawns, etc., and with the intentional consumption of *Aphanizomenon flos-aquae* (AFA)-based Blue-Green Algae Supplements (BGAS), shown to be contaminated with cyanotoxins, specifically MCs, (Schaeffer *et al.*, 1999; Gilroy *et al.*, 2000; Lawrence *et al.*, 2001a). MCs are the most commonly found group of cyclic heptapeptide cyanotoxins with molecular weights ranging between 900 and 1100 Dalton (Da), represent more than 80 structural variants differing in the two variable L-amino acids (Meriluoto and Spoof, 2008).

The *in vivo* and *in vitro* toxicity of MCs is primarily governed by the potent inhibition of serine/threonine-specific protein phosphatases (PPs) (MacKintosh *et al.*, 1990; Eriksson *et al.*, 1990a; MacKintosh and MacKintosh, 1994; Toivola *et al.*, 1994), specifically PP1, PP2A, PP4 and PP5 (Hastie *et al.*, 2005). As a consequence of PP inhibition numerous cellular proteins e.g. intermediate filaments, are hyperphosphorylated, thereby leading to the collapse of the cytoskeleton and loss of cellular integrity (Eriksson *et al.*, 1989; Ghosh *et al.*, 1995; Batista *et al.*, 2003). Cellular necrosis and apoptosis is observed in a dose- and time dependent manner, whereby apoptosis is observed at lower concentrations than overt necrosis (Mankiewicz *et al.*, 2001; Fladmark *et al.*, 2002; Gehringer, 2004; Fu *et al.*, 2005; Weng *et al.*, 2007). Due to their structure and size MCs do not readily penetrate the cell membrane via simple diffusion but rather require the presence of multi-specific organic anion transporting polypeptides (rodent Oatp / human OATP) for active uptake (Runnegar *et al.*, 1991; Fischer *et al.*, 2005; Komatsu *et al.*, 2007; Monks *et al.*, 2007). It is thus not surprising that co-incubation of MCs, e.g. MC-LR, with the known Oatp/OATP substrates choline, taurocholate (TC) and bromosulfophthalein (BSP) reduces the uptake of MC *in vitro* (Runnegar *et al.*, 1995; Fischer *et al.*, 2005; Komatsu *et al.*, 2007; Monks *et al.*, 2007). Moreover, a knock-out mouse lacking expression of mOatp1b2 in the liver presented with no overt liver pathology when exposed to MC-LR (Lu *et al.*, 2008), while the wild-type counterpart showed the typical hepatotoxicity observed in *i.p.* or oral MC-LR *in vivo* exposure experiments with mice (Ito *et al.*, 2000). Oatp/OATP are primarily expressed in enterocytes, hepatocytes and renal epithelial cells (Kullak-Ublick *et al.*, 1994; Bergwerk *et al.*, 1996; Abe *et al.*, 1999; König *et al.*, 2000a; König *et al.*, 2000b; Kullak-Ublick *et al.*, 2004; Mikkaichi *et al.*, 2004; Sai *et al.*, 2006; Naud *et al.*, 2007; Tani *et al.*, 2008) as well as in heart, lung, spleen, pancreas, brain and the blood-brain-barrier (BBB) (Hagenbuch and Meier, 2003; Hagenbuch and Meier, 2004). Consequently, the systemic distribution of MCs is governed by the degree of blood perfusion and the type and expression level of Oatp/OATP present in a given organ. Due to the first-pass effect, high blood perfusion and high expression level of multiple Oatp/OATP types, MCs are often characterized as hepatotoxins, although other organs may also be affected. Indeed, Fischer and Dietrich (2000) treated carp (*Cyprinus carpio*) with a single dose of 400 µg/kg bw MC-LR and demonstrated pathological changes in the hepatopancreas and kidney as well as the presence of MC-LR in several organs including the brain 48 h post toxin application (Fischer and Dietrich, 2000). Immunoblotting of brain homogenates with anti-MC-LR antibody revealed a band with molecular weight of approximately 38 kDa, corresponding to the catalytic subunits of PP1 and PP2A (37 kDa), thus corroborating the interaction of MC-LR with PPs in the brain. A comparable study with mice demonstrated the rapid appearance of MC-LR in the brain, 45 min and 60 min after the initial intraperitoneal (i.p.) or peroral (p.o.) administration, respectively (Meriluoto *et al.*, 1990; Nishiwaki *et al.*, 1994). As well as detection of MC, pathological changes of the brain anatomy were observed in five day old progeny of mice which had been treated with a toxic extract of *Microcystis aeruginosa*. Of these neonatal mice 10% showed a reduced brain size and some presented with extensive pathology in the outer region of the hippocampus (Falconer *et al.*, 1988). The above data suggest that as a consequence of the high blood perfusion of the brain, significant amounts of MC could reach the brain across the BBB and induce brain pathology. Indeed, the above hypothesis is supported by the tragic events in February 1996, where 131 patients at a hemodialysis clinic in Caruaru, Brazil, employing poorly treated drinking water and overly used dialysis cartridges, were exposed intravenously to varying concentrations of MC-congeners (MC-YR/-LR/-AR) (Jochimsen *et al.*, 1998; Carmichael *et al.*, 2001; Azevedo *et al.*, 2002; Soares *et al.*, 2006). Of these MC-exposed patients (mean approximate value of 19.5 µg $MC_{equiv.}$/l in dialysis water (Carmichael *et al.*, 2001)), 89% developed immediate signs of neurotoxicity (e.g. dizziness, tinnitus, vertigo, headache, vomiting, nausea, mild deafness, visual disturbance and blindness) with a later onset of overt hepatotoxicity and finally succumbed to multi-organ failure. Of the patients exposed 76 patients died within 10-weeks of initial intravenous exposure (Pouria *et al.*, 1998).

Different Oatp/OATP types appear to have varying affinities for MCs. Indeed, uptake of MC-LR *in vitro* via liver-specific OATP1B1, 1B3, Oatp1b2 (mouse, rat), Oatp1d1 (skate) as well as the more widely distributed (kidney, liver, BBB) OATP1A2 have previously been described (Fischer *et al.*, 2005; Komatsu *et al.*, 2007; Meier-Abt *et al.*, 2007; Monks *et al.*, 2007; Lu *et al.*, 2008). Thus the observed liver failure in the Caruaru incident was most likely a direct consequence of the liver-specific uptake of MCs via OATP, e.g. OATP1B1 and 1B3 (Fischer *et al.*, 2005; Komatsu *et al.*, 2007; Monks *et al.*, 2007), whereas the immediate neurotoxicity may be explained by OATP-mediated transport, e.g. OATP1A2 (Fischer *et al.*, 2005), of MCs across the BBB. Indeed, OATP1A2 is highly expressed in endothelial cells of the BBB, epithelial cells of the blood-

2. Manuscript I

cerebrospinal-fluid-barrier (BCFB) and in the cell membrane of human neurons (Kullak-Ublick *et al.*, 1995; Gao *et al.*, 2000; Gao *et al.*, 2005; Lee *et al.*, 2005; Nies, 2007).

Of the mouse Oatps, mOatp1a1, mOatp1a4, mOatp1a5, and mOatp1a6 belong to the same OATP1A family, i.e. having greater than 60% amino acid sequence identity, as the human OATP1A2 (Hagenbuch and Meier, 2004). Similarly the mOatp1b2 belongs to the same OATP1B family as the rat rOatp1b2 and the human OATP1B1 and OATP1B3, while the mOatp1c1 belongs to the OATP1C family with the human OATP1C1. However, in contrast to the known transporting capabilities of the human OATP (1A2, 1B1, 1B3) (Fischer *et al.*, 2005; Komatsu *et al.*, 2007; Monks *et al.*, 2007), it is currently not known whether or not the latter also applies to human OATP1C1. The most recent comparison of the skate Oatp1d1, demonstrated to be able to transport MC-LR at a low level, with other OATPS of the OATP family tree (Hagenbuch and Meier, 2004), suggests that the skate Oatp1d1 is an evolutionarily ancient precursor of the mammalian-liver OATP1B family, however exerts the highest degree of homology (50.4% amino acid sequence identity) with the human OATP1C1 of the OATP1C family (Meier-Abt *et al.*, 2007). Based on the degree of evolutionary conservation of mOatp, as denoted by the high amino acid sequence identity with human OATP demonstrated to being capable of transporting MC, it was assumed that mouse Oatp have similar MC transporting capabilities. Indeed, the latter assumption is at least partially corroborated by Lu et al (Lu *et al.*, 2008), who demonstrated lack of acute MC-induced hepatotoxicity in mOatp1b2- knock-out mice. Moreover, as mouse Oatp1a1, 1a4, 1a5, 1c1, 2b1, and 3a1 were demonstrate to be expressed (mRNA level) in the mouse brains (Hagenbuch and Meier, 2004), the question was raised whether one or more of these mOatps, i.e. 1a1, 1a5, 1c1, and 3a1, could be involved in transporting MC into neuronal cells. Mouse Oatps1b2 and 6d1 were included in the analysis as mOatp1b2 was assumed to be primarily expressed in the liver (Cheng *et al.*, 2005; Lu *et al.*, 2008) while no knowledge on brain expression was available for mOatp 6d1.

In order to test the hypothesis that MCs are taken up actively into neuronal cells in a mOatp-associated and MC-congener-dependent manner, the presence of mOatp in primary murine whole brain cells (mWBC) was verified at the mRNA and protein level. MC transport and neuronal toxicity was tested with three MC congeners (MC-LR, -LW, -LF) with/without co-incubation with the OATP/Oatp substrates TC and BSP. Uptake of MCs was demonstrated indirectly via cytotoxicity measurements and directly via MC-Western blot analysis, protein phosphatase inhibition determination, and via MC-specific immunocytochemistry in mWBC cultures exposed to MCs.

2. Manuscript I

Materials and methods

Chemicals and reagents

All chemicals, unless otherwise stated were of the highest analytical grade commercially available. Individual MC-congeners (MC-LR, -LW and -LF) were obtained from Alexis Biochemicals, Lausen, Switzerland; Okadaic acid (OA) from Sigma-Aldrich, Taufkirchen, Germany.

Ham's F12 medium (F12), minimal essential medium (MEM), Iscove's modified DMEM (IMDM), fetal bovine serum (FBS), penicillin/streptomycin and G-418-sulphate (Geneticin) were purchased from PAA Laboratories, Pasching, Austria and poly-L-lysine, trypsin, trypsin inhibitor and bovine serum albumin (BSA) were obtained from Sigma-Aldrich, Taufkirchen, Germany.

[^{32}P]-ATP and Amersham ECL Plus Western Blotting detection reagents were purchased from GE Healthcare, Munich, Germany; Adenosine 5'-triphosphate disodium salt, phosphorylase b from rabbit muscle and phosphorylase kinase from rabbit muscle from Sigma-Aldrich, Taufkirchen, Germany and protease inhibitor cocktail set III was obtained from Calbiochem, San Diego, U.S.A.

For reverse transcription-Polymerase Chain Reaction (RT-PCR), M-MuLV RT (1000u), 5x reaction buffer (supplied with M-MuLV RT), Oligo(dT)$_{18}$ primer (100 µM), random hexamer primer (100 µM), dNTP Mix (10mM), Ribonuclease Inhibitor (RiboLock, 2500u) and for PCR the 2x PCR Master Mix were obtained from Fermentas, St. Leon-Rot, Germany. Primer pairs were purchased from MWG-Biotech, Martinsried, Germany.

Animals

Specific pathogen-free Balb/c mice were obtained from The Jackson Laboratory, Bar Harbor, U.S.A. and held at the animal facility, University of Konstanz, Germany. Sacrifice and organ removal was carried out in accordance with the German Animal Protection Law, approved by the relevant German authority, the Regierungspräsidium in Freiburg, Germany (registry number: T-07 05).

Isolation, cell culture and characterization of primary murine WBC

Six-to seven-day-old pups were decapitated and whole brains were immediately removed, cut into small pieces and then trypsinized in HIB solution (120 mM NaCl, 5 mM KCl, 25 mM HEPES, 9.1 mM Glucose) containing 2.5 g/l trypsin for 12 min at 37°C. Trypsin inhibitor (3.75 g/l) was added and mWBC were centrifuged for 5 min at 300 x g. The supernatant was discarded and the resulting pellet was re-suspended in culture medium (1:1 (v/v) IMDM/F12, supplemented with 10% heat inactivated FBS and 1% penicillin-streptomycin). Cells were gently triturated with a 1.5 inch 21 gauge needle and then filtered through a 100 µm nylon mesh. The dissociated mWBC were seeded in poly-L-lysine (50 mg/l) coated 6 and 96 well plates (Greiner Bio-One, Frickenhausen, Germany)

2. Manuscript I

and 8 well chamber slides (BioCoat; BD, Heidelberg, Germany) at a density of 1.6×10^6 cells/ml and cultured at 37°C and 5% CO_2. Culture media was renewed every 72 h.

To characterize the primary murine whole brain cells (mWBC), Western blots were carried out using brain cell-type specific antibodies kindly provided by Prof. Leist (Doerenkamp-Zbinden Chair of Alternative in-vitro Methods, University of Konstanz). Briefly, following 8 days of *in vitro* cultivation, the presence of neurons, astrocytes, microglia, endothelial cells and fibroblasts in the mWBC culture, as well as in homogenates of whole brains of six-day-old pups, were assessed via Western blot using antibodies to anti-class III β-tubulin (1:1000; Covance, Emeryville, U.S.A Figure 2.1 A), anti-glial fibrillary acidic protein (1:1000; Sigma-Aldrich, Taufkirchen, Germany; Figure 2.1 B), anti-F4/80 (1:1000; AbD serotec, Martinsried, Germany; Figure 2.1 C) and anti-vimentin (1:1000; developmental studies hybridoma bank, university of Iowa, U.S.A). Positive signals were observed in mWBC for neurons, astrocytes and microglia (Figure 2.1 A-C), vimentin, representing endothelial cells and fibroblast type cells, could not be detected (figure not shown), corroborating earlier findings in mWBC (Bologa *et al.*, 1983).

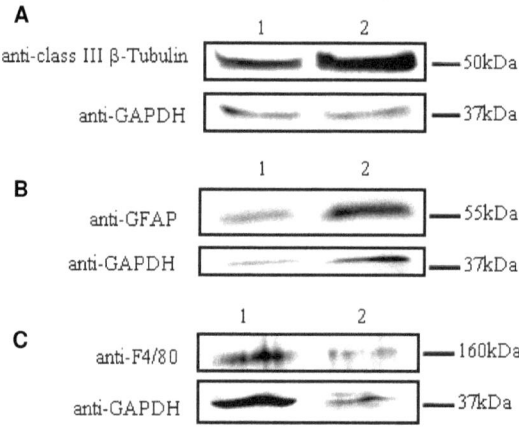

Figure 2.1:
Characterization of mWBC using cell type-specific antibodies and Western blot. (A) Neuron-specific class III β-tubulin antibody in mWBC (lane 1, 15 µg total protein) and murine brain homogenate (lane 2, 15 µg total protein). (B) Astrocyte-specific glial fibrillary acidic protein (GFAP) antibody in mWBC (lane 1, 20 µg total protein) and murine brain homogenate (lane 2, 20 µg total protein). (C) MCA497G antibody against murine microglial surface protein F4/80 in mWBC (lane 1, 40 µg total protein) and murine brain homogenate (lane 2, 20 µg total protein).

Toxin exposure

mWBC were cultured for 7-8 days (70-80% confluency) and incubated for 48 h with the toxins. MC-LR, -LW and –LF were dissolved in 75% methanol (MeOH), TC, BSP and OA in water.

Incubation with 70 nM OA served as positive control (0% survival) in cytotoxicity studies. mWBC incubated with culture medium only served as negative control (100% survival). The concentration of MeOH never exceeded 0.5%, which served as solvent control in all experiments. No differences in viability, condition or growth rate could be identified between solvent and negative control. For co-incubation studies a single MC-congener was incubated with TC (ACROS, Geel, Belgium) or BSP (Sigma-Aldrich, Taufkirchen, Germany) in excess (50 and 500 µM) for 48 h. TC and BSP alone demonstrated no reduction of cell viability at the concentrations used (data not shown).

Determination of MC uptake in mWBC via Western blot analysis (WB) and immunocytochemistry (ICC)

Western blot: The medium of MC-exposed mWBC was removed, cells were washed 3 times with phosphate-buffered saline (PBS) and homogenized in 60 µl ice-cold extraction buffer (10 mM triethanolamine (Tris)-base, pH 7.5, 140 mM NaCl, 5 mM EDTA, 0.1% (v/v) Triton X-100, 1x protease inhibitor cocktail). To remove cell membrane fragments, the lysate was centrifuged for 5 min at 16000 x g and 4°C. The protein content was determined by the method of Bradford (Bradford, 1976) (Bio-Rad Protein Assay; Bio-Rad, Munich, Germany) and equal protein amounts (30 µg/lane) were separated using a 10% sodium dodecyl sulfate-polyacrylamide gel electrophoresis (SDS-PAGE) according to the method of Laemmli (Laemmli, 1970) at constant 200 volts. Following electrophoresis, proteins were transferred onto nitrocellulose membranes (Whatman, Dassel, Germany) at 300 mA for 90 min as previously described (Tobwin *et al.*, 1979). Membranes were incubated with blocking buffer (TTBS – Tris-buffered saline with Tween 20 (100 mM Tris-HCl, 0.9% (w/v) NaCl, pH 7.6, 0.1% (w/v) Tween 20) containing 1% bovine serum albumine (BSA) for 30 min at RT and incubated with polyclonal rabbit anti-MCLR#2 (1:500) (Mikhailov *et al.*, 2001), monoclonal mouse anti-ADDA (1:700) (clone AD4G2, Alexis Biochemicals, Lausen, Switzerland), monoclonal mouse anti-PP1α (1:500) (Sigma-Aldrich, Taufkirchen, Germany), monoclonal mouse anti-PP2A/C (1:1000) (Upstate, Temecula, U.S.A.) and monoclonal mouse anti-GAPDH (1:30000) (Sigma-Aldrich, Taufkirchen, Germany) for 16 h at 4°C. Secondary antibodies were horseradish peroxidase (HRP) -conjugated mouse anti-rabbit- (1:50000) (Sigma-Aldrich, Taufkirchen, Germany) and goat anti-mouse (1:20000) (Sigma-Aldrich, Taufkirchen, Germany) antibodies. Immunopositive bands were visualised via ECL substrate according to the manufactures recommendations and the resulting chemiluminescence signal was detected using Fujifilm LAS-1000 (FUJIFILM Electronic Imaging, Kleve, Germany). For re-probing nitrocellulose membranes, blots were stripped using 50 ml of 50°C stripping buffer (50 mM Tris-base, 2% (w/v) SDS, pH 6.8, 100 mM β-mercaptoethanol) for 30 min under gentle shaking.

Immunocytochemistry: The medium of MC-exposed mWBC was removed, cells were washed 3 times with PBS and fixed in -20°C cold ethanol/acetone (1:1) for 2 min on ice. The fixation buffer was discarded and the chamber slides were air dried for 30min and stored at -20°C. For the detection of MC, mWBC were re-hydrated in PBS and incubated in blocking buffer containing 1% BSA for 30 min. Rabbit anti-MCLR#2 antibody (1:500) (Mikhailov et al., 2001) was applied in a humidified atmosphere and incubated for 60min at RT. The fluorochrome-conjugated secondary antibody (1:500) goat anti-rabbit-Alexa488 (λ_{max}: 495 nm; Invitrogen, Karlsruhe, Germany), was added and samples were incubated for 60 min at RT. For cytoskeleton detection a concentration of 700 nM TRITC-phalloidin (λ_{max}: 547 nm; Sigma-Aldrich, Taufkirchen, Germany) was used to stain actin filaments and nuclei were counterstained with 2.5 µM Hoechst 33342 (λ_{max}: 352 nm; Invitrogen, Karlsruhe, Germany). Finally, culture slides were mounted with Fluorescent Mounting Medium (DAKO, Hamburg, Germany) and visualized using a con-focal microscope (LSM 510 META, Zeiss, Göttingen, Germany).

Determination of MC-congener specific cytotoxic effects in mWBC

Cytotoxicity assay: mWBC cell viability was assessed by determination of 3-(4,5 dimethylthiazol-2-yl)-2,5-diphenyl tetrazolium bromide (MTT) reduction according to Mosmann (Mosmann, 1983). Briefly, after 48 h toxin exposure, 25 µl MTT solution (5 mg/ml, ACROS, Geel, Belgium) was added to each well and incubated for 90 min at 37°C. After incubation the medium was gently removed and 100 µl solubilization buffer (95% (v/v) isopropanol, 5% (v/v) formic acid) was added to each well, shaken carefully for at least 15 min to re-dissolve the formazan product. Absorption was measured at 550 nm using a microtiter plate reader (Tecan, Crailsheim, Germany). The test was carried out independently three times in duplicates for each congener.

Radioactive protein phosphatase inhibition assay (rPPIA)

Phosphorylation of phosphorylase b with [^{32}P]-ATP and the protein phosphatase inhibition assay were performed as described by Fischer and Dietrich (Fischer and Dietrich, 2000a).

In a preliminary step, cell extracts of untreated mWBC were taken through a dilution series with extraction buffer (see immunoblotting) in order to determine the linear range of dephosphorylation of the [^{32}P]-ATP-labeled substrate. Protein concentration was determined according to the method of Bradford (Bradford, 1976). In the final assay 30 µg of total protein in a volume of 20 µl per sample was employed. mWBC extracts were added to 20 µl of protein phosphatase assay buffer (consisting of 0.1 mM EDTA, 20 mM imidazole-HCl, pH 7.63, 1 mg/ml BSA, 0.1% (v/v) β-mercaptoethanol final concentration). The reaction was started by adding 20 µl [^{32}P]-phosphorylase a in solubilization buffer (50 mM Tris-HCl, pH 7.0, 0.1 mM EDTA, 15 mM caffeine, 0.1% (v/v) β-

2. Manuscript I

mercaptoethanol). After incubation for 5 min at 30°C the reaction was stopped by addition of 180 ml ice-cold 20% (w/v) trichloroacetic acid and cooling on ice for at least 10 min. Subsequently the samples were centrifuged for 5 min at 12000 x g. In order to extract free [^{32}P] 180 µl of the resulting supernatants were mixed with 200 µl of acid phosphate buffer (1.25 mM KH_2PO_4 in 0.5 M H_2SO_4), 500 µl isobutanol/heptane (1:1) and 100µl ammoniumheptamolybdate (5% (w/v)) by vigorously vortexing. Radioactivity was counted in a liquid scintillation counter (LS 6500; Beckman Coulter, Krefeld, Germany) after mixing 300 µl of the [^{32}P]-containing solvent layer with 1 ml of scintillation cocktail (Ready Safe; Beckman Coulter, Krefeld, Germany). Total ser/thr-specific protein phosphatase activity of MC- and OA- (positive control) exposed mWBC was calculated by determining the percentage loss of radioactivity from untreated cells (negative control = 100% activity). Each sample was analyzed three times in duplicate. The degree of protein phosphatase inhibition expressed as activity of the corresponding control was normalized to the corresponding number of viable cells. This was achieved via division of the mean protein phosphatase inhibition with the corresponding median of cytotoxicity (MTT assay) observed.

Determination of mOatp transporters in mWBC
RNA extraction: Total RNA was isolated from mWBC using the RNeasy Mini Kit (Qiagen, Hilden, Germany) according to the manufacturer's instructions. Column-bound total RNA was eluted with 60 µl RNase-free water and purity was determined by the quotient 260 nm/280 nm of optical density (OD). Stably transfected HEK293-OATP1B3 (Komatsu *et al.*, 2007) (cultured in MEM supplemented with 10% FBS, 1% penicillin/streptomycin and 400 µg/ml G418-sulphate at 37°C and 5% CO_2) were employed as an internal OATP positive control. Culture medium was removed every 72 h until cells were used for the experiment (80-90% confluence).

Reverse transcription-polymerase chain reaction (RT-PCR)
Each reaction was carried out in a final volume of 21 µl. Therefore 6.8 µl of total RNA (1.3 µg), 3.2 µl RNase-free water, 0.5 µl random hexamer primer and 0.5 µl oligo(dT) were incubated for 5 min at 70°C following 5 sec at 25°C and chilled for at least 1 min on ice. Reaction buffer (4 µl), 2 µl dNTP mix, 0.5 µl ribonuclease inhibitor and 1.5 µl RNase-free water were added (5 min, 25°C). Finally, 2 µl of Reverse Transcriptase (RT) or RNase-free water (for RT negative controls) was incubated for 10 min at 25°C, 60 min at 37°C following 10 min at 70°C.
PCR for mOatp/OATP was performed in 25 µl reaction mixtures according to Mühlhardt (Mühlhardt, 2000) with minor modifications. Primers were designed (Primer3 software; Rozen and Skaletzky, 2000) re-hydrated with RNase-free water to the appropriate volumes according to MWB-Biotech data sheets (100 pmol/µl), gently mixed and shaken over night at 4°C. The

2. Manuscript I

sequences of primer pairs used in this assay are shown in Table 2.1. Briefly, amplification was subsequently carried out by mixing 2 µl of cDNA product with 12.5 µl of 2x PCR Master Mix solution, 4 µl of the primer pair mix (0.5 µM forward- and 0.5 µM reverse primer) and 6.5 µl of RNase-free water. Human OATP1B3 was used as a positive control. PCR reaction was performed in a thermocycler (Primus 96 Plus; MWG-Biotech, Martinsried, Germany) under the following conditions: initial denaturing for 3 min at 94°C, 30 cycles of denaturing for 1 min at 94°C, annealing for 1 min at 58°C and elongation for 1 min at 72°C followed by an additional extension for 7 min at 72°C.

Agarose gel electrophoresis of PCR products

Each PCR product (20 µl) was separated by gel electrophoresis on a 3% agarose gel buffered in TBE (10.8 g/l Tris-base, 5.5 g/l boric acid, 4 ml (v/v) 0.5 M EDTA ph 8.0). DNA ladder (GeneRuler 100bp; Fermentas, St. Leon-Rot, Germany) was employed to determine the PCR product size. Gels were stained by ethidium bromide, visualized under UV light (302 nm), photographed (Polaroid camera and Polaroid 667 films) and then scanned for documentation.

Purification and sequence confirmation of PCR products

Following 3% agarose gel separation of PCR products, bands were excised and the PCR products purified via microcolumn elution (QIAquick gel extraction kit; Qiagen, Hilden, Germany). Purified PCR products were sent to GATC (Konstanz, Germany) for DNA sequencing. Sequence results received were aligned (http://blast.ncbi.nlm.nih.gov/bl2seq/wblast2.cgi) with the corresponding known mOATP/OATP1B3 sequences.

Table 2.1: Primer sequences used for PCR

Gene (Protein)	Primer	Sequence 5′ → 3′	Product size (base pairs)	Degree of alignment (BLAST)
Slco1c1 (Oatp1c1)	Sense	gtaggggattccagctcctc	204	100%
	Antisense	gcataatgagcccaaaagga		
Slco1a1 (Oatp1a1)	Sense	atccagtgtgtggggacaat	235	100%
	Antisense	atggctgcgagtgagaagat		
Slco1a5 (Oatp1a5)	Sense	gcacagagaaaaagccaagg	166	100%
	Antisense	ctccaggtatttgggcaaga		
Slco3a1 (Oatp3a1)	Sense	gccttttggtgaagaagctg	275	100%
	Antisense	gaagcaggctgacaggtagg		
Slco1b2 (Oatp1b2)	Sense	ttcaccacaacaatggccta	194	100%
	Antisense	ttttccccacagacaggttc		
Slco6d1 (Oatp6d1)	Sense	gaagcaggctcaggtggtag	249	n.d.
	Antisense	acgaccgctaaaaacgacag		
SLCO1B3 (OATP1B3)	Sense	gggtgaatgcccaagagata	168	100%
	Antisense	attgactggaaacccattgc		

n.d.: not detectable due to limited band visibility and thus PCR product extraction.

Investigation of mOatp1b2 protein expression in mWBC via Western blot analysis

For analysis of mOatp1b2 via WB, mWBC were cultured and homogenized as described above. Briefly, the mWBC homogenate was centrifuged for 40 min at 16000 x g and 4°C in order to obtain the crude membrane fraction. The pellet was resuspended and protein concentration was determined by the method of Bradford (Bradford, 1976) (Bio-Rad Protein Assay; Bio-Rad, Munich, Germany). Proteins were separated by SDS-PAGE, blotted on a nitrocellulose membrane and incubated with an antibody against mOatp1b2 (Lu *et al.*, 2008) kindly provided by Prof. Klaassen (Department of Pharmacology, Toxicology and Therapeutics, University of Kansas Medical Center, Kansas, USA).

Statistics

Statistical analysis was performed by one-way analysis of variance (ANOVA) with Dunnett's post test, Bonferroni or Newman-Keuls multiple comparison test, where appropriate, using GraphPad Prism 4.03, $P<0.05$ (*), $P<0.01$ (**) and $P<0.001$ (***).

Results

Uptake of MC-LR into mWBC

A concentration-dependent uptake of MC-LR was observed in mWBC exposed to MC-LR for 48 h (Figure 2.2 A). MC-LR-positive bands ranged between ~39 and ~28 kDa. The ~39 kDa band corresponded to the catalytic subunits of PP1 (37,5 kDa) and PP2A (36 kDa), as confirmed via immunoblotting using antibodies against PP1α and PP2A/C (Figure 2.2 B), yet in addition containing a covalently bound 1 kDa MC. The latter was also confirmed by MC-incubation studies using purified PP1 and PP2A, which provided a comparable ~39 kDa band (Figure 2.2 C).

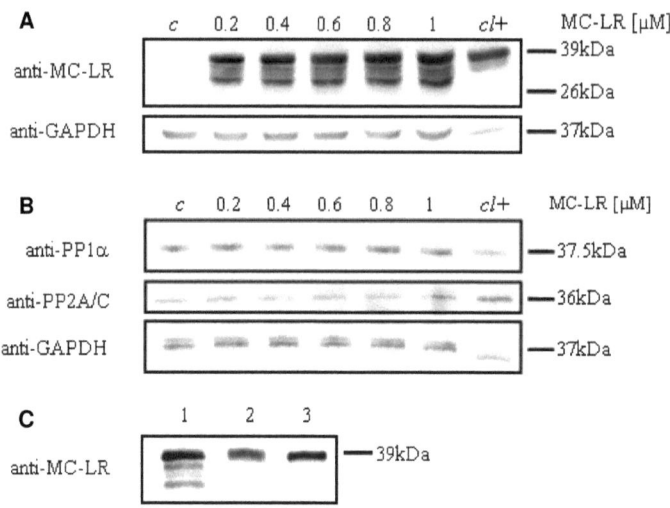

Figure 2.2:
(A) Determination of intracellular MC-LR after 48h exposure of mWBC to a range of MC-LR concentrations. Western blots using MC-antibody MCLR#2 (Mikhailov et al., 2001). c: control mWBC, cl+: 1μM MC-LR added to mWBC cell lysate; (B) Determination of intracellular PP1α and PP2A/C expression in mWBC exposed to MC-LR. Western blots using PP1a and PP2A/C antibodies. (C) Determination of PP-associated MC-LR after 48h exposure of mWBC. Lane 1: mWBC following exposure to 1μM MC-LR, lane 2: purified rabbit muscle PP1 (New England Biolabs, Germany) incubated with 1μM MC-LR *in vitro*, lane 3: purified human erythrocyte PP2A (Promega, Germany) incubated with 1μM MC-LR.

Intracellular localization and cytotoxic effects in mWBC

Incubation (48 h) of mWBC with 0, 0.6 and 5.0 μM MC-LR and subsequent immunofluorescence analyses using con-focal microscopy demonstrated the presence of MC-LR in the cytosol and the nuclei of mWBC (Figure 2.3 B and C). Loss of structural integrity of the cytoskeleton was MC-LR concentration-dependent, as shown by the congregation of actin filaments around the nuclei (Figure 2.3 B and C). Loss of cytoskeleton integrity also coincided with protein phosphatase inhibition (Figure 2.4) but not directly to cytotoxicity as determined via MTT assay (Figure 2.5). However, complete congregation of actin filaments around the nuclei (Figure 2.3 C) as observed following incubation with 5 μM MC-LR corresponded to approximately 40% reduction in cell viability, as determined via MTT.

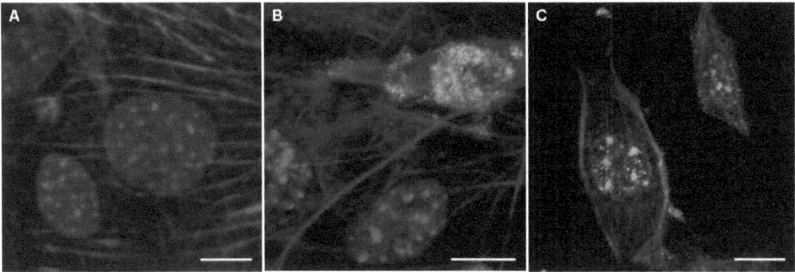

Figure 2.3:
Intracellular MC-LR localization and associated cytoskeleton abnormalities in mWBC exposed to MC-LR for 48h. Immunolabeling of MC-LR (green: MCLR#2-Alexa488), actin filaments (red: TRITC-Phalloidin) and nucleus (blue: Hoechst 33342); (A) control mWBC; (B) mWBC exposed to 0.6µM MC-LR; (C) mWBC exposed to 5µM MC-LR. Scale bar 10µm.

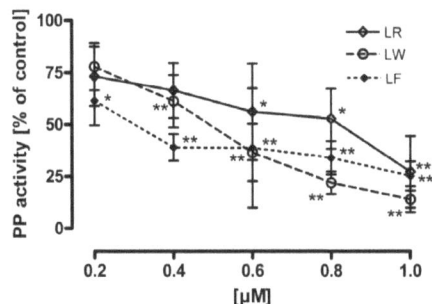

Figure 2.4:
Protein phosphatase activity in mWBC after treatment with MC-congeners –LR, -LW, or –LF normalized to median of cytotoxicity. Values represent mean ± SD of three independent experiments and are expressed as percentage of control. Statistics: one-way ANOVA with Dunnett's post-test ($P<0.05$ "*"; $P<0.01$ "**").

MC-congener specific cytotoxic effect

Exposure of mWBC for 48 h to MC-LF, -LW and -LR resulted in significant inhibition of protein-phosphatase in mWBC (Figure 2.4) at ≥200 nM, ≥400 nM and ≥600 nM, respectively. Protein-phosphatase inhibition corresponded with the observed cytotoxicity at the highest concentrations (Figure 2.5). Indeed, while exposure of mWBC to 5 µM MC-LF resulted in a complete loss of viable cells, the same concentrations of MC-LW and –LR resulted only in a 33% and 54% reduction of cell viability, respectively.

2. Manuscript I

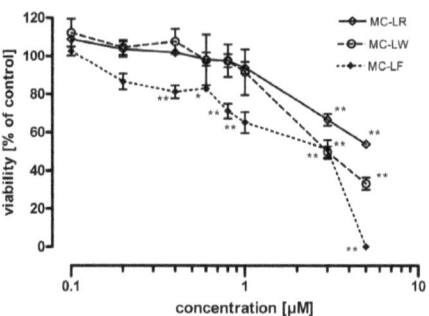

Figure 2.5:
Cytotoxicity in mWBC following 48h exposure to MC-LR, -LW, and –LF (conc. range 0.1 - 5µM), determined via MTT. Viability values represent mean ± SEM of three independent experiments and are expressed as percentage of control. Statistics: one-way ANOVA with Dunnett's post-test (P<0.05 "*"; P<0.01 "**").

Determination of mOatp-associated transport of MC

To further evaluate the role of bile acid transporters in the active uptake of different MC congeners into mWBC, the presence of various mOatps was examined at the mRNA level. mWBC were screened for six mOatps (Oatp1c1, 1a5, 3a1, 1a1, 1b2 and 6d1) and resulted in single positive bands for five tested mOatps (Oatp1c1, 1a5, 3a1, 1a1 and 1b2) with 166 – 275 bp corresponding to the expected size of amplified products with the primers employed (Figure 2.6). Water and RT negative controls did not show any contamination/bands (data not shown). The amplified products were sequenced to confirm the detection of the visualized mOatp fragments. While five of the six mOatp RT-PCR products could be confirmed via sequencing and alignment analysis (Table 2.1), there was insufficient DNA extractable for mOatp6d1 (Slco6d1) to allow sequence confirmation.

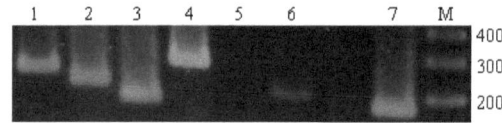

Figure 2.6:
RT-PCR of Oatp-mRNA expression in mWBC. Lane 1: Oatp1a1; lane 2: Oatp1c1; lane 3: Oatp1a5; lane 4: Oatp3a1; lane 5: Oatp6d1; lane 6: Oatp1b2; lane 7: OATP1B3 (positive control: stably transfected HEK293-OATP1B3). M: Gene ruler (bp).

In addition, mOatp1b2 was detectable in mWBC, whole brain and liver homogenates (positive control, (Lu et al., 2008)) at the protein level (Figure 2.7) with a molecular weight of 70 – 80 kDa (www.uniprot.org; UniProtKB/Swiss-ProtQ9JJL3-1(SO1B2_Mouse), mOatp1b2 known molecular weight 76,729 kDa).

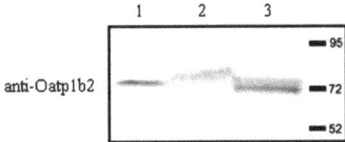

Figure 2.7:
Western Blot of mOatp1b2 in mWBC crude membrane fractions using an anti- Oatp1b2 polyclonal antibody (Lu *et al.*, 2008). Lane 1: mWBC (40 µg); lane 2: murine brain homogenate (40 µg); lane 3: murine liver homogenate (positive control, 40 µg).

In order to corroborate the association of mOatp with the active uptake of MC congeners in mWBC, co-incubation studies were carried out employing excess concentrations of the OATP/Oatp substrates TC and BSP (Jacquemin *et al.*, 1994; Runnegar *et al.*, 1995; Kanai *et al.*, 1996; Fischer *et al.*, 2005; Monks *et al.*, 2007). The viability of mWBC exposed to MCs were compared to mWBC exposed to the combination of 50 and 500 µM TC (Figure 2.8 B, C, E, F, H and I) or 50 µM BSP (Figure 2.8 A, D and G) and the corresponding MCs. Co-incubation of TC and BSP with MCs generally resulted in a reduction of observed cytotoxicity. For MC-LR this reduced cytotoxicity was observed at ≥5 µM MC-LR only, whereas reduced cytotoxicity was already observed at 1 and 3 µM for MC-LW and –LF, respectively. Moreover, while the reduction of cytotoxicity in MC-LW and – LF exposed mWBC co-incubated with 50 µM BSP was limited, the corresponding experiments with 50 and 500 µM TC demonstrated a much greater reduction in MC-mediated cytotoxicity. Provided that MC are primarily transported via mOatps, the data as presented here suggest better MC transport-inhibition by TC than by BSP, despite that the degree of reduction in MC-mediated cytotoxicity was comparable in the 50 and 500 µM TC co-incubation experiments.

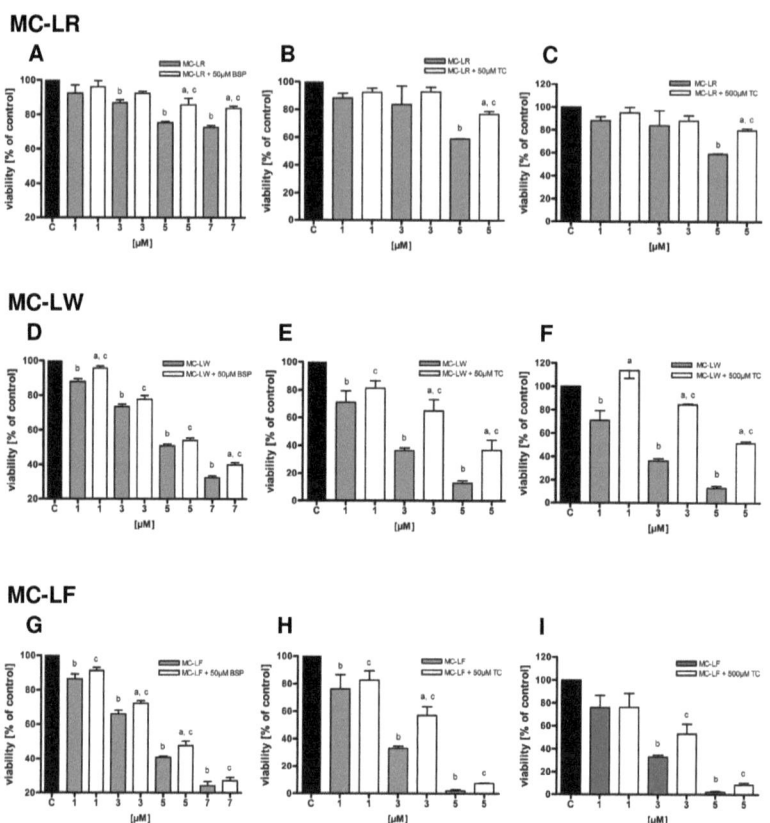

Figure 2.8:
Cytotoxic effects of MC-LR (A, B, C), -LW (D, E, F), -LF (G, H, I) in mWBC co-incubated with/without the Oatp substrates taurocholate (TC: 50 and 500 µM) and bromosulfophthalein (BSP: 50 µM). Control mWBC (C: black columns) represent 100% cell viability, MC-congener treated mWBC (grey columns) and MC-congener + TC/BSP co-incubated mWBC (white columns) following 48h exposure. Values represent mean ± SEM of three independent experiments. Statistics: One-way ANOVA with Newman-Keuls multiple comparison test (comparison of a: MC vs. MC + BSP/TC; b: MC vs. control; c: MC + BSP/TC vs. control, all at $P<0.05$).

TC-induced reduction of MC transport was also corroborated by Western blots demonstrating smaller protein bands positive for MC-LW in cells co-incubated with TC than in cells exposed to MC-LW only (Figure 2.9). Moreover, the width of the MC-LW positive protein bands decreased in a MC-LW concentration (1.0 – 0.05 µM MC-LW) dependent manner, being smallest in cells exposed to 50 nM MC-LW and 500 µM TC.

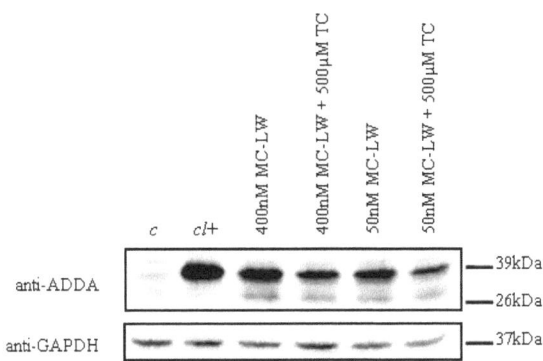

Figure 2.9:
Oatp-dependent uptake of MC-LW in mWBC following 48h co-incubation with/without the Oatp substrate taurocholate (TC: 500 µM). Western blot using anti-ADDA antibody (clone AD4G2) for MC-LW adduct detection.

Discussion

Contamination of human water sources with the cyanobacterial toxin MC has been shown to induce hepato- and neurotoxic effects in patients of a dialysis clinic in Caruaru, Brazil (Pouria et al., 1998). The early onset of neurotoxicity observed could be explained by the uptake of MCs via OATPS e.g. OATP1A2 which is highly expressed in endothelial cells of the BBB, epithelial cells of the BCFB and in the membrane of human neurons (Kullak-Ublick et al., 1995; Gao et al., 2000; Fischer et al., 2005; Gao et al., 2005; Lee et al., 2005; Nies, 2007). However, neurotoxicity can only evolve if MCs are actively and effectively transported, e.g. via MC-transport competent Oatp/OATP, or possibly other yet unknown transporters, expressed in brain cells. Transport of MC-LR has been demonstrated for human OATP1A2, 1B1, 1B3, rat/mouseOatp1b2 and scateOatp1d1 (Fischer et al., 2005; Komatsu et al., 2007; Meier-Abt et al., 2007; Monks et al., 2007; Lu et al., 2008) but not for the human OATP2B1 and rat Oatp 1a1 and 1a4 (Fischer et al., 2005), thus demonstrating that not all Oatp/OATP are capable of transporting MC-LR. The latter however also raised the question whether different MC congeners are transported by different Oatp/OATP albeit with varying efficiency, thereby suggesting that an Oatp/OATP not transporting MC-LR does not exclude its capability of transporting other MC congeners.

The results presented in this study, clearly demonstrate the expression of at least five mOatps in primary murine whole brain cells (Figure 2.6) at the mRNA level. The different band intensities as observed in Figure 2.6 most likely result from varying binding efficiencies of the respective primers generated to the respective mWBC mOatp mRNAs. The detection of the liver specific MC-LR

2. Manuscript I

transporter Oatp1b2 (mouse, rat) (Fischer *et al.*, 2005; Lu *et al.*, 2008) in liver and brain homogenates as well as in mWBC at the protein level (Figure 2.7), suggest the functional expression of mOatp1b2 in mouse brains and mWBC. As Oatp1b2 (mouse, rat) was demonstrated to actively transport MC-LR the observed MC-mediated cytotoxicity and PP inhibition in mWBC may be at least partially due to mOatp1b2 mediated MC transport into mWBC. However, other mOatps may also be involved in the transport of MCs, since Oatp1b2 knock-out mice have shown to develop no overt hepatotoxicity, albeit detectable MC-LR were observed in liver homogenates (Lu *et al.*, 2008). Incubation of mWBC with the three different MC congeners MC-LR, -LW- and LF resulted in congener-dependent toxicity as demonstrated by the observed concentration-dependent cytotoxicity (Figure 2.5) and protein phosphatase inhibition (Figure 2.4). Moreover, the observed variant cytotoxcity with the three different MC congeners (Figure 2.8) in conjunction with the findings by Lu et al (Lu *et al.*, 2008) suggest that although mOatp1b2 efficiently transports MC-LR, mOatp1b2 mediated MC transport may not be as important in mouse brains and that other mOatp, expressed at higher levels than mOatp1b2, as well as MC congeners other than MC-LR may be more important for MC-mediated neurotoxicity. However, only specific expression of individual mOatp and MC-transport analyses, as discussed below, will allow determination of the contribution of each individual mOatp expressed in mWBC and mouse brains in the observed MC-congener specific transport.

Intracellular localization of MCs was proven via immunocytochemistry (Figure 2.3) and Western blotting of mWBC cell homogenates (Figure 2.2 and Figure 2.9), thus conclusively demonstrating transport of MCs into mWBC. mOatp-associated MC transport, is also supported by the fact that uptake of MCs into mWBC, as indicated by the observed cytotoxicity (Figure 2.8) and MC-Western blotting of cell homogenates (Figure 2.9), was reduced upon co-incubation of mWBC with the OATP/Oatp substrates TC and BSP. Both 50 µM BPS as well as 50 and 500 µM TC reduced uptake of MC congeners, albeit TC with a much greater efficacy than BSP. Provided that the observed MC cytotoxicity is mediated via mOatp dependent transport, the observed differences in reduction of MC-congener dependent cytotoxicity by BSP and TC may stem from varying affinities of the three MC-congeners, TC and BSP for the mOatp detected in the mWBC employed. However, only specific expression of these mOatp, e.g. in *Xenopus* oocytes (Fischer *et al.*, 2005), HeLa (Monks *et al.*, 2007) or HEK293 cells (Komatsu *et al.*, 2007) and subsequent kinetic analyses of MC-congener dependent transport would allow deduction whether the mOatps in mWBC are primarily responsible for MC transport.

The data of this study thus suggest that MC-LF and MC-LW have a much higher potential for inducing neurotoxicity in mice than MC-LR. Consequently, based on the high similarities of mOatp with human OATP, a more thorough investigation of different MC congeners appears important for proper assessment of microcystin toxicity, including neurotoxicity, and cancer risk in humans, rather than relying on MC-LR risk assessment (WHO, 1998).

Acknowledgements

This study was supported by the International Research Training Group 1331 (IRTG 1331). We would like to thank, Dr. Elisa May (Bio Imaging Centre, University of Konstanz, Germany) for the introduction to con-focal microscopy, Prof. Dietrich Keppler (Division of Tumor Biochemistry, German Cancer Research Centre, Heidelberg, Germany) for providing stably transfected HEK293-OATP1B3 and Alexandra Heussner for technical support. Special thanks are due to Dr. Bernhard Ernst for great support with the rPPIA assay and for his contributions to this project. The editorial review of the manuscript by Dr. Evelyn O'Brien is gratefully acknowledged. Finally we would like to thank Oliver Okle for technical assistance.

3. Manuscript II

ORGANIC ANION TRANSPORTING POLYPEPTIDES EXPRESSED IN PRIMARY MURINE NEURONAL CELLS MEDIATE MICROCYSTIN CONGENER DEPENDENT UPTAKE

Feurstein D., Kleinteich J., Stemmer K., Dietrich D.R.

Human and Environmental Toxicology, University of Konstanz, Konstanz, Germany

Submitted for publication in Environmental Health Perspectives

Abstract

Contamination of natural waters by toxic cyanobacteria is a growing worldwide problem, resulting in serious water pollution and human health hazards. Microcystins (MCs) represent a group of >80 cyclic heptapeptides, inducing cytotoxicity via specific protein phosphatase (PP) inhibition at equimolar concentrations (comparable toxidynamics). Due to MCs structure and size, active uptake into cells occurs via organic anion transporting polypeptides (OATP/Oatp), as confirmed for liver specific human OATP1B1, 1B3, mouse mOatp1b2, scate Oatp1d1 and more widely distributed OATP1A2 expressed e.g. at the blood-brain-barrier. Tissue and cell-type specific expression of OATP/Oatp transporters and specific transport of MC congeners (toxicokinetics) therefore appear prerequisite for reported hepato-, nephro- and neurotoxic effects in humans and other species upon MC-exposure. Beyond MC-LR induced hepatotoxicity, the effects of other MC congeners especially neuronal uptake and toxicity are unknown. This study therefore examined the expression mOatps and the uptake of MC-LR, -LW and –LF in primary murine neurons. Twelve mOatp mRNAs were found, while the presence of mOatp1b2, a known MC-LR transporter, was also confirmed at the protein-level. Functionality of mOatps was confirmed with OATP/Oatp substrates taurocholate (TC) and estrone sulfate (ES). Uptake of ES, but not TC, was competitively inhibited by MC-LF and to a lesser extent by MC-LW and -LR. Neuronal uptake of MCs was confirmed via western-blot analysis and via a significant decrease of neuronal PP activity. Above data suggest a mOatp dependent uptake of specific MC congeners into neurons, thus corroborating earlier assumptions of the neurotoxic potential of microcystins.

Keywords: cyanobacteria, microcystin congeners, organic anion transporting polypeptides, neurotoxicity

3. Manuscript II

Introduction

Cyanobacteria are abundant in marine, brackish and freshwaters. During the past decades, cyanobacterial poisonings of aquatic and terrestrial animals as well as humans (Kuiper-Goodman *et al.*, 1999; Stewart *et al.*, 2008) have been reported worldwide especially in conjunction with cyanobacterial blooms in water reservoirs, rivers, lakes and ponds used for drinking water or recreational purposes or as water resources for livestock. Current research suggests that climate change, specifically global warming, will promote the incidence and severity of cyanobacterial mass occurrences (Carmichael, 2008; Paerl and Huisman, 2008; Paul, 2008), thereby increasing the potential threat posed by cyanobacterial toxins to human health and livestock (WHO, 1999).

Microcystin (MC), the most common freshwater cyanotoxin, represents a group of cyclic heptapeptides encompassing more than 80 structural variants (Meriluoto and Spoof, 2008) with molecular weights ranging between 900 and 1100 Da (Zurawell *et al.*, 2005). Their inherent capability for inhibiting serine/threonine-specific protein phosphatases (PPs) e.g. PP1, PP2A, PP4 and PP5 via covalent binding to the PP catalytic subunit (Honkanen *et al.*, 1990; MacKintosh *et al.*, 1990; Hastie *et al.*, 2005) results, once taken up into the cell, in a disruption of cellular phosphorylation/dephosphorylation homeostasis. This in turn leads to several downstream responses e.g. disintegration of cytoskeletal structure, inhibition of gluconeogenesis, and enhanced glycolysis (Falconer and Yeung, 1992; Batista *et al.*, 2003), frequently with later onset of apoptosis and necrosis (Fladmark *et al.*, 1999; Fu *et al.*, 2005; Komatsu *et al.*, 2007). However, in order to exert the MC-specific PP inhibition, sufficient concentrations of MC must enter the cell. Cellular uptake of MCs has been demonstrated to occur exclusively via an active transport, while passive trans-membrane diffusion can be excluded (Eriksson *et al.*, 1990; Fischer *et al.*, 2005; Komatsu *et al.*, 2007). Consequently, pathological changes following MC intoxications are restricted to organs, tissues and cells capable of actively transporting MC from the blood into the cell. Indeed, active trans-membrane transport of MCs, is mediated by specific organic anion transporting polypeptides (human OATP / rodent Oatp).

Currently 80 carriers of the OATP/Oatp superfamily in at least 13 different species have been identified with 36 representing members in human, mouse and rat (Hagenbuch and Meier, 2004; Hagenbuch, 2007) expressed either ubiquitously (e.g. OATP1A2) or with a tissue/organ specific expression, (e.g. OATP1B1 and 1B3 in the liver) (Mikkaichi *et al.*, 2004). However, not all OATPs/Oatps are capable of transporting MCs (Fischer *et al.*, 2005) and different OATPs/Oatps appear to have largely differing affinities and capacities for MC congeners (Monks *et al.*, 2007; Feurstein *et al.*, 2009). Thus, the latter highlights the fact that OATPs/Oatps capable of transporting MC need to be functionally expressed in a tissue/organ or cell type such that MCs can exert a cytotoxic effect. Indeed, this was demonstrated convincingly with Oatp1b2 knockout mice, being resistant to the overt MC-LR hepatotoxicity observed in corresponding wild-type mice (Lu et al., 2008). In consequence, the often quoted hepato- and nephrotoxicity of MCs, is the result of a hepatic first-pass- and subsequent renal elimination-effect in organs having a high level of functionally expressed OATPs/Oatps capable of MC-transport. More recently, several OATPs/Oatps were described in the blood-brain-barrier (BBB), the blood-cerebrospinal-fluid-barrier (BCSFB), in human gliomas and in glia cells (Gao et al., 2000; Bronger et al., 2005; Huber et al., 2007; Westholm et al., 2009). Therefore it may be assumed that MCs are able to enter the brain and to exert neurotoxic effects. Indeed, 116 (89%) of 131 patients of a hemodialysis unit in Caruaru, Brazil, accidentally exposed to MCs via dialysis water (specifically MC-LR, -YR and –AR) (Pouria et al., 1998; Carmichael et al., 2001; Azevedo et al., 2002), presented with acute symptoms of neurotoxicity e.g. deafness, tinnitus and reversible blindness. Subsequently 100 patients developed liver failure, of which 76 died (Pouria et al., 1998; Carmichael et al., 2001). Furthermore, a reduction in brain sizes was reported in progeny of Swiss Albino mice exposed to cyanobacterial bloom extract containing MCs (Carmichael et al., 1988), thus suggesting that MCs have an effect on the brain. Whether the observed neurological effects in the Caruaru patients stem from an effect of MCs on the endothelium of the BBB with subsequent *in-situ* ischemia and inflammatory reactions, or a direct uptake of MCs via OATPs of the BBB endothelium (Cecchelli et al., 2007) and OATPs of astrocytes, microglia and/or neurons, remain to be resolved. Recently, acute MC neurotoxicity was demonstrated in primary murine whole brain cells (Feurstein et al., 2009). This study, while confirming the presence of murine Oatps (mOatps) and MC congener dependent cytotoxicity, however did not distinguish between the various cell types affected (e.g. astrocytes, microglia or neurons).

Thus, whilst the literature strongly suggests the presence of OATPs/Oatps capable of MC transport in the BBB, the expression of OATPs/Oatps in neurons and MC-congener neurotoxicity still remains elusive. In view of the scarcity of human primary neurons, mouse primary neurons were employed to determine the identity of mOatps expressed, to confirm mOatp-mediated MC congener-specific uptake and determine MC congener specific inhibition of neuronal protein phosphatases.

Materials and methods

Materials

^{3}[H]taurocholate (TC) (170.2 Giga Becquerel (GBq)/mmol) and ^{3}[H]estrone sulfate, ammonium salt (ES) (2120 GBq/mmol) were from PerkinElmer Life and Analytical Sciences (Bosten, USA). MC congeners were from Alexis Biochemicals, Lausen, Switzerland and reverse transcriptase-polymerase chain reaction (RT-PCR) chemicals were from Fermentas, St. Leon-Roth, Germany. All chemicals and antibodies, unless otherwise stated, were from Sigma-Aldrich (Taufkirchen, Germany). Cell culture media and reagents were from PAA Laboratories (Pasching, Austria).

Isolation and cultivation of primary neurons

Specific pathogen-free Balb/c mice were obtained from The Jackson Laboratory (Bar Harbor, U.S.A) and held at the animal facility University of Konstanz, Germany under standard conditions. Sacrifice and organ removal was carried out in accordance with the German Animal Protection Law, approved by the Regierungspräsidium in Freiburg, Germany (registry number: T-07 05). Primary murine cerebellar granule neurons (Holtzman et al., 1987; Skaper et al., 1996) were isolated and cultured as described previously (Volbracht et al., 1999) with minor modifications. Briefly, cerebelli from six-to seven-day-old mouse pups were removed, separated into single cells and seeded in poly-L-lysine (50 mg/L) coated plates at a density of 2.0×10^6 cells/mL. Neurons were cultured in basal medium (Eagle's) (BME), supplemented with 10% fetal calf serum, 20 mM potassiumchloride and 1% penicillin-streptomycin at 37°C in a humidified 5% carbondioxide (CO_2) incubator. After 24 hr, culture medium was renewed containing 10 µM cytosine arabinoside and cells were grown for three days without further medium exchange. Subsequently neurons were harvested for mOatp screening or MC exposure experiments (see below).

Determination of mOatp mRNA expression in primary neurons

Total RNA was isolated using the RNeasy Mini Kit (Qiagen, Hilden, Germany) according to manufacturer's instructions. RT-PCR was carried out in a final volume of 20 µL containing 10.5 µL total RNA (0.5 µg) diluted in RNase-free water, 1 µL random hexamer primer (50 µM) and 1 µL oligo(dT) (50 µM). The mixture was incubated at 70°C for 10 min to promote primer annealing. Subsequently, 5x reaction buffer (4 µL), 1 µL desoxynucleotid-triphosphate mix (10 mM) and 2 µL dithiothreitol (DTT, 0.1 M) were added and incubated at 42°C for 2 min. Finally, cDNA synthesis was carried out using 0.5 µL SuperScript II (100U, Invitrogen Karlsruhe, Germany) for 50 min at 42°C followed by an inactivation for 15 min at 70°C. PCR was performed in 20 µL reaction mixtures according to Mühlhardt (Mühlhardt 2002) with minor modifications. Specific primers for mOatps (Table 3.1) were designed using Primer3 software (Rozen and Skaletzky 2000). mOatp amplification was carried out by mixing 1 µL of cDNA product and cycling conditions for all reactions as follows: 4.5 min at 95°C, 35 cycles of 30 sec at 95°C, 1 min at 59°C, 1min at 72°C, 1 min at 95°C and 1 min at 55°C.

3. Manuscript II

Table 3.1: Sequences of the primers used in this study.

Protein (Gene)	Primer	Sequence 5' → 3'	Product size (base pairs)	PCR Signal
Hprt1 (housekeeping)	forward	agcttgctggtgaaaagga	186	yes
	reverse	ttgcgctcatcttaggcttt		
Oatp1a1 (Slco1a1)	forward	tggggagaaaaatgtccttg	169	yes
	reverse	gcagctgcaattttgaaaca		
Oatp1a4 (Slco1a4)	forward	accagcatccccttttctt	221	yes
	reverse	aaggcattgacctggatcac		
Oatp1a5 (Slco1a5)	forward	acatggcattctgcctatca	293	yes
	reverse	gcagctgcaattttgaaaca		
Oatp1a6 (Slco1a6)	forward	aatgccaaagaggagaagca	215	yes
	reverse	actgcctttgctgtggagat		
Oatp1b2 (Slco1b2)	forward	ttcaccacaacaatggccta	194	yes
	reverse	ttttccccacagacaggttc		
Oatp1c1 (Slco1c1)	forward	gtagggattccagctcctc	204	yes
	reverse	gcataatgagcccaaaagga		
Oatp2a1 (Slco2a1)	forward	cctatgctcagggagacgag	201	yes
	reverse	gcagtagtcccagctgaagg		
Oatp2b1 (Slco2b1)	forward	gactatggctccagcctctg	320	yes
	reverse	tggggtctttggagtcaag		
Oatp3a1 (Slco3a1)	forward	ggattcgagggatgcagtta	190	yes
	reverse	aaggcagatttgcagcttgt		
Oatp4a1 (Slco4a1)	forward	ggtggcttcctggtaaacaa	194	yes
	reverse	caagctgcctttaggtccag		
Oatp4c1 (Slco4c1)	forward	tggctctttgcttggtcttt	309	yes
	reverse	tgcaaagctcgatgtcaatc		
Oatp5a1 (Slco5a1)	forward	ctgctctgcaaaagggattc	290	yes
	reverse	tgaggtatccagccctcatc		
Oatp6b1 (Slco6b1)	forward	caaattggccacaaccttct	233	n.d.
	reverse	ccttagaacctgggcaatca		
Oatp6c1 (Slco6c1)	forward	accagcgaaggtgtaccaac	256	n.d.
	reverse	ggtcagcatcagtgggttct		
Oatp6d1 (Slco6d1)	forward	gaagcaggctcaggtggtag	249	n.d.
	reverse	acgaccgctaaaaacgacag		

Western-blot (WB) analysis for mOatp1b2 and MC-LR detection

Neurons were homogenized for WB and visualization of mOatp1b2, MC-LR and glyceraldehyde-3-phosphate-dehydrogenase (GAPDH), as previously described (Feurstein *et al.*, 2009). Briefly, protein samples were separated by sodium dodecyl sulfate polyacylamid gel electrophoresis, transferred to a nitrocellulose membrane. mOatp1b2 (polyclonal rabbit anti-mOatp1b2, 1:1000, kindly provided by Prof. Klaassen, University of Kansas Medical Center, Kansas, USA), MC-LR (polyclonal rabbit anti-MCLR#2, 1:500 (Mikhailov *et al.*, 2001)), and GAPDH as house-keeping control protein (monoclonal mouse anti-GAPDH 1:30000) were used as primary antibodies and incubated for 16 hr at 4°C. Secondary horseradish peroxidase-conjugated goat anti-rabbit- (1:160000) and rabbit anti-mouse (1:80000) antibodies were incubated for 1 hr at room temperature (RT). Immunoreactive bands were detected by enhanced chemiluminescence (GE Healthcare, Munich, Germany).

3. Manuscript II

Immunocytochemistry for mOatp1b2 detection

Cells were, fixed in 4% paraformaldehyd for 15 min at RT and subsequently incubated with blocking buffer (phosphate buffered saline (PBS) containing 1% bovine serum albumin (BSA)) for 1 hr at RT. Polyclonal rabbit anti-mOatp1b2 (1:800) was applied and slides incubated for 16 hr at 4°C. Secondary fluorochrome-conjugated goat anti-rabbit-Alexa Fluor 647 antibody (Invitrogen, Karlsruhe, Germany) was applied in a 1:1000 dilution and incubated for 1 hr at RT. Nuclei were counterstained with 2.5 µM Hoechst 33342 (Invitrogen, Karlsruhe, Germany) for 10 min at RT. Finally, slides were visualized using an Axiovert 200M microscope (Zeiss, Göttingen, Germany).

TC and ES uptake studies

The transport capacity of neuronal mOatps was assessed via ^{3}[H]TC and ^{3}[H]ES saturation kinetics. ^{3}[H]TC and ^{3}[H]ES are specific substrates for and thus transported by a variety of OATPs/Oatps (van Montfoort et al., 2003; Konig et al., 2006). Saturation kinetic experiments were initiated by addition of 500 µL BME medium pre-warmed at 37°C containing varying concentrations of either ^{3}[H]TC (3.7 kilo (k)Bq/mL) or ^{3}[H]ES (3.7 kBq/mL) and unlabeled TC and ES, respectively. Following 30 min exposure (37°C) uptake was stopped by removing the medium and a subsequent washing step using 3x 1.5 ml of ice cold PBS per well. Neurons were solubilised with 500 µL of 0.2 normal sodium hydroxid per well (15 min) and the solution neutralized with 500 µL of 0.2 normal hydrogen chloride (HCl). Solubilised neurons were mixed with 10 mL of scintillation cocktail (Ready Safe, Beckman Coulter, Krefeld, Germany) and radioactivity was determined in a liquid scintillation counter (LS 6500; Beckman Coulter, Krefeld, Germany). To normalize radioactivity per mg protein, protein concentration was determined according to Bradford (Bradford, 1976). Individual samples were analyzed in triplicates (N=3), each with technical duplicates.

MC congener dependent inhibition of TC/ES uptake

Competitive uptake experiments were carried out as described above, albeit using 500 µL BME medium containing either 7 µM mixture of ^{3}[H]TC (3.7 kBq/mL) and TC or a 7 µM mixture of ^{3}[H]ES (3.7 kBq/mL) and ES in absence or presence of single MC congeners (7 µM). To normalize radioactivity per mg protein, protein concentration was determined as described above. Individual samples were analyzed in triplicates (N=3), each with technical duplicates.

PP activity after exposure to single MC congeners

Cells were exposed to varying concentrations of single MC congeners for 48 hr and solubilised in 50 µL enzyme solution buffer (0.08 M Tris-HCl, pH 7.0, 0.19 mM ethylene glycol tetraacetic acid 20 mM DTT (in 0.01 M sodium acetate, pH 5.2), 10 mM manganese chlorid ($MnCl_2$)). Protein concentrations were determined via the method of Bradford (Bradford, 1976). Subsequently, 20 µL of each sample (containing 10 µg protein) was transferred to a 96-well plate and an equal volume of water was added. To determine PP activity, 200 µL of freshly prepared and 37°C pre-warmed substrate solution containing reaction buffer (0.25 M Tris-HCl, pH 8.1, 10mM $MnCl_2$, 0.2 M magnesium chloride 5 mg/mL BSA), 60 mM p-nitrophenyl phosphate (pNPP, Acros, U.S.A.) and 20 mM DTT were added and immediately measured at 405 nanometer (TECAN infinite M200, TECAN, Craislheim, Germany) (0 min value). Total PP activity of MC exposed neurons was calculated after incubation for 90 min at 37°C by determining the percentage loss of pNPP substrate turnover from untreated cells (100% PP activity). Individual samples were analyzed in triplicates (N=3), each with technical duplicates.

Data Analysis

Kinetic parameters were estimated using Michaelis-Menten kinetics:

$v = V_{max} * S / (K_m + S)$,

whereby v is the uptake rate [pico (p)mol/mg*min] of the substrates, V_{max} is the maximum uptake rate [pmol/mg*min], S the substrate concentration in the medium [µM] and K_m the Michaelis constant [µM]. All data were expressed as means ± S.E. and fitted to the equation by nonlinear regression (GraphPad Prism 4.03). Statistical analysis was performed by one-way analysis of variance (ANOVA) with Dunnett's post test using GraphPad Prism 4.03, P<0.05 (*), P<0.01 (**).

Results

Determination of mOatp expression in primary murine neurons

Twelve of the 15 mOatps were screened for and showed up as individual single positive bands (Figure 3.1) corresponding to the expected size of amplified products (mOatp1a1, 1a4, 1a5, 1a6, 1b2, 1c1, 2a1, 2b1, 3a1, 4a1, 4c1 and 5a1). None of the three members of the mOatp6 family (mOatp6b1, 6c1 and 6d1) were found. Water and RT negative controls were examined and indicated no contamination/bands (data not shown).

3. Manuscript II

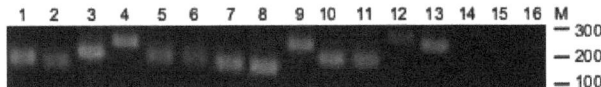

Figure 3.1:
RT-PCR of mOatp mRNA expression in primary neurons. 1: Hprt1 (housekeeping gen), 2: mOatp1a1, 3: mOatp1a4, 4: mOatp1a5, 5: mOatp1a6, 6: mOatp1b2, 7: mOatp1c1, 8: mOatp2a1, 9: mOatp2b1, 10: mOatp3a1, 11: mOatp 4a1, 12: mOatp4c1, 13: mOatp5a1, 14: mOatp6b1, 15: mOatp6c1, 16: mOatp6d1, M: DNA ladder (bp).

mOatp1b2 protein was determined using crude membrane fractions of primary murine neurons and mouse liver homogenates (positive control; Lu *et al.*, 2008)). WB analysis (Figure 3.2 C) demonstrated immunopositive bands with molecular weights of approximately 80 kDa (mOatp1b2: 76.729 kDa (UniProtKB/Swiss-ProtQ9JJL3-1(SO1B2_Mouse) and 60 kDa. In addition, the localization and expression of mOatp1b2 in neurons was assessed via immunocytochemistry and revealed immunopositive staining in membranes of the perikaryon as well as in neurites of cultured cells (Figure 3.2 B). Unspecific binding of the fluorochrome-conjugated secondary antibody to neurons (negative control) was negative and thus can be excluded (Figure 3.2 A).

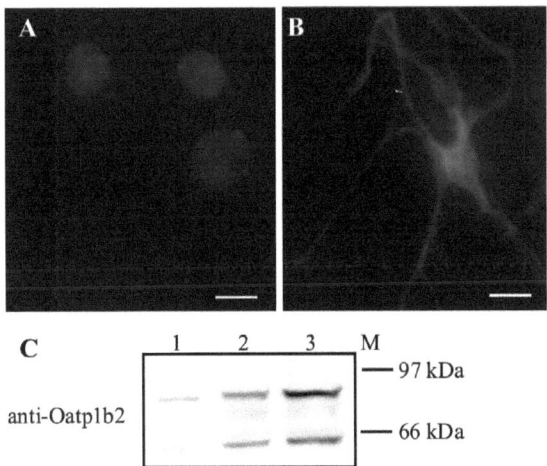

Figure 3.2:
Immunodetection of mOatp1b2 in cultured neurons. (A) neurons treated with Hoechst (blue) and fluorochrome-conjugated secondary antibody only (negative control). (B) neurons treated with Hoechst, rabbit anti-mOatp1b2 and fluorochrome-conjugated secondary antibody (red). Scale bar: 10 µm. (C) WB analysis of mOatp1b2 in neuronal- (lane 1, 15 µg) and liver crude membrane fractions (positive control: lane 2, 15 µg; lane 3, 30 µg). M: protein marker.

Kinetics of TC and ES uptake

Saturable uptake was observed in neurons for both substrates, TC and ES (Figure 3.3 A and B) with increasing substrate concentrations. Computer estimated Michaelis-Menten kinetic analysis suggested K_m and V_{max} values for TC and ES of 47.72 ± 20.87 µM and 20.79 ± 13.93 pmol/mg*min and of 12.32 ± 1.66 µM and 6.01 ± 1.34 pmol/mg*min, respectively.

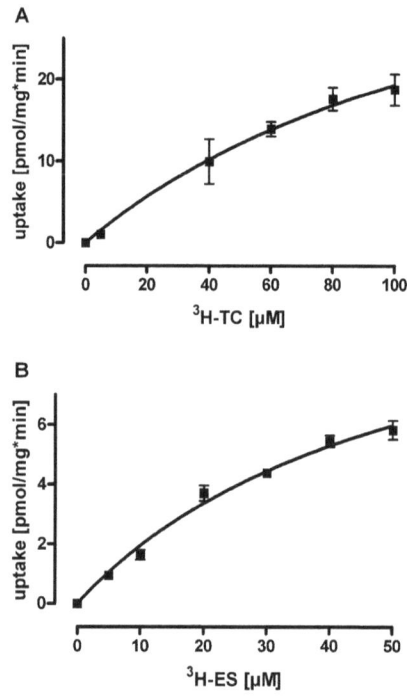

Figure 3.3:
Uptake kinetics of (A) ^{3}[H]TC and (B) ^{3}[H]ES in neurons. Individual samples were analyzed in triplicates (N=3), each with technical duplicates and given as mean values ± S.E.M. Note difference in scales of x- and y-axes in A and B.

Competitive inhibition of TC/ES uptake by MC congeners

TC and ES uptake in neurons was competitively inhibited by MC-LR, -LW and –LF. Irrespective of the MC-congener, TC uptake was inhibited by 20% (Figure 3.4 A). In contrast while MC-LR or MC-LW inhibited ES uptake by 30%, a 45% reduction of ES uptake was observed with MC-LF (Figure 3.4 B) thus suggesting MC congener specific competition with ES for mOatps in neurons.

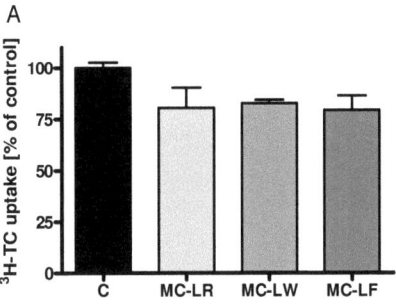

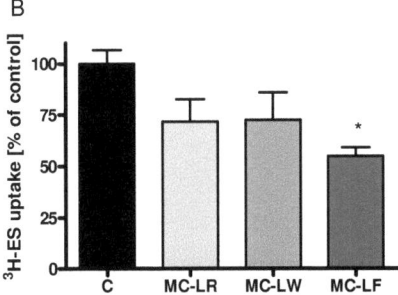

Figure 3.4:
Inhibitory effects of MCs on the active uptake of (A) ^{3}[H]TC and (B) ^{3}[H]ES in neurons. Cells were exposed to ^{3}[H]TC and TC (7 µM) or ^{3}[H]ES and ES (7 µM) in absence or presence of single MC congeners (7 µM). The inhibitory effect of individual MCs was calculated by determining the percentage loss of co-incubated versus TC/ES exposed neurons (100% uptake). Individual samples were analyzed in triplicates (N=3), each with technical duplicates and given as mean values ± S.E.M.; $P<0.05$ (*).

Intracellular PP inhibition by MCs

Transport of the three MC congeners into neurons was indirectly confirmed using a colorimertic PP inhibition assay. At low MC concentrations (0.31 µM – 1.25 µM) the inhibitory effect of all MC congeners was comparable and resulted in a 20% reduction of total PP activity when compared to control (Figure 3.5). At 2.5 µM MC-LR, -LW and –LF total PP activity was reduced by 25%, 30% and 60%, respectively, while 5 µM MC-LF reduced total PP activity by 65%.

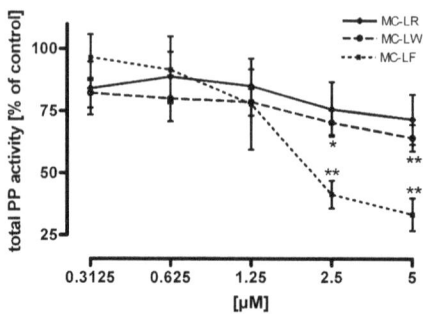

Figure 3.5:
Total PP activity determination of primary neurons after MC-LR, -LW, and -LF exposure. Cells were treated with varying concentration of single MC congeners for 48 h. Individual samples were analyzed in triplicates (N=3), each with technical duplicates and given as mean values ± S.E.M. and are expressed as percentage of control (untreated cells: 100% PP activity); P<0.05 (*), P<0.01 (**).

Confirmation of intracellular MC-LR

Confirmation of MC-LR uptake into murine neurons was achieved by MC-LR WB analysis (Figure 3.6), demonstrating a concentration-dependent uptake of MC-LR. The single MC-LR-positive band had an approximate molecular weight of 39 kDa, thus corresponding to the expected size of the catalytic subunits of PP1 (37.5 kDa) and PP2A (36 kDa) with the covalently bound MC-LR (1 kDa).

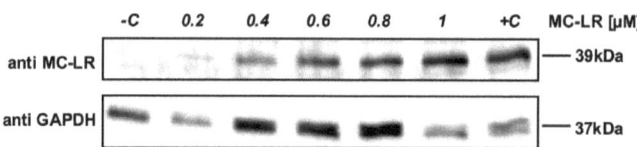

Figure 3.6:
Intracellular- and concentration-dependent detection of MC-LR via WB analysis. Neurons were exposed to varying MC-LR concentrations for 48 h and subsequent immunolabeled using an MC-LR specific antibody. -c: untreated neurons, +c: 1µM MC-LR added to a neuronal homogenate.

3. Manuscript II

Discussion

Eriksson et al. (Eriksson et al., 1987) demonstrated >20 years ago that a cyclic peptide toxin isolated from the cyanobacterium *Microcystis aeruginosa* presented with a cell-type specific cytotoxicity. Three years later the same authors showed that a multispecific bile acid transport system is involved in the uptake of MC-LR into primary rat hepatocytes (Eriksson et al., 1990). Only more recently was it possible to conclusively demonstrate that OATPs/Oatps are responsible for transporting MC-LR (Fischer et al., 2005; Komatsu et al., 2007; Meier-Abt et al., 2007; Lu et al., 2008). However, it was also shown that not all OATPs/Oatps are capable of transporting MC-LR (Fischer et al., 2005) and that OATPs/Oatps may decisively differ in their ability to transport the structurally variant MC-congeners (Monks et al., 2007; Feurstein et al., 2009; Fischer et al., in press). Thus the potential neurotoxicity of individual MC congeners largely depends on the functional expression of OATPs/Oatps at the BBB/BCSFB (Bronger et al., 2005; Huber et al., 2007; Westholm et al., 2009) and in the neuronal cell membrane. A further prerequisite is the MC congener transporting capability of the OATPs/Oatps expressed in these locations. Using the mouse as a surrogate system for humans, it was possible to demonstrate the mRNA of 12 mOatps in primary murine neurons, thus suggesting the presence of 12 functional mOatps in mouse neurons (Figure 3.1). The latter finding was supported by the fact that mOatp1b2 was not only detected at the mRNA level but also at the protein level (Figure 2 B and C). Although WB analysis of mOatp1b2 provide for two positive bands at approximately 80 and 60 kDa, the upper band (~80 kDa) represents mOatp1b2 with a molecular weight of 76.729 kDa (UniProtKB/Swiss-ProtQ9JJL3-1 (SO1B2_Mouse)) while the lower band (~60 kDa) most likely represents either a deglycosilated variant or a fragment of mOatp1b2. Indeed, double bands of in WB analyses of OATP1B1/OATP1B3/mOatp1b2 proteins have previously been described (Komatsu et al., 2007; Zaher et al., 2008).

Although it was not possible to directly demonstrate the functional expression of mOatps in mouse neurons and specific mOatp mediated transport of MCs, support was gleaned from three separate lines of indirect evidence:

Covalently bound MC-LR was detectable in WB of mouse neuron homogenate preparations following exposure to MC-LR (Figure 3.6). Moreover, the bands detected corresponded to the molecular weights of the catalytic subunits of PP1 and PP2A with covalently bound MC-LR and were comparable to those published previously by Lu et al. (Lu et al., 2008) and Feurstein et al (Feurstein et al., 2009). Using both MC-LR treated wild-type and Oatp1b2$^{-/-}$ mice, Lu et al. (Lu et al., 2008) was able to demonstrate that mOatp1b2 is an important carrier of MC-LR but also highlighted the fact that other mOatps capable for transporting MC-LR are expressed in the murine liver with variant MC-LR transporting capabilities.

3. Manuscript II

Variant mOatp transporting capabilities were also observed in the mouse neuron uptake kinetic experiments when using the well-characterized OATP/Oatp substrates TC and ES, demonstrating an approximately 4- and 3-fold lower K_m and V_{max} value, respectively, for ES than for TC (Figure 3.3 A and B). As all of the OATPs/Oatps known to transport MCLR are also established transporters of TC and ES (van Montfoort et al., 2003; Konig et al., 2006), it is not surprising that co-incubations of TC or ES with equimolar concentrations of MC-LR, -LW, and -LF appeared to competitively inhibit uptake of TC and ES (Figure 3.4 A and B). However, although significant for MC-LF only, the overall competitive effect exerted by all MC congeners appeared to be strongest upon competition with ES, while this effect was at best marginal for TC. As the Km values for TC and ES determined in the murine neurons (Figure 3.3) represent a summary Km value determined over all TC/ES transporting mOatps, the MC uptake competition experiments suggest that only a few of the mOatps found in murine neuron, i.e. those transporting ES, are also capable of transporting MCs. Provided the latter assumption holds true, mOatps transporting TC with higher affinity and capacity than ES, should be less inhibited than mOatps with similar affinity and capacity for TC and ES. Although little is known about the affinity and capacity for ES, TC or MCs of mOatps, the comparison with the human orthologues does demonstrate that OATP1A2 has comparable Km values for TC and ES, while OATP1B1 and possibly OATP1B3, 2B1 and 4A1 had lower Km values for ES than for TC (van Montfoort et al., 2003; Konig et al., 2006). The latter would suggest that the murine orthologue of the human OATP1A2 would be the most likely candidate for efficient MC transport in murine neurons. However, as of yet the murine orthologue of the human OATP1A2 is unknown or absent, suggesting that one of the other mOatps identified is responsible for MC transport. In addition, the competition experiment data with ES also demonstrated that MC-LF appeared to compete more strongly with ES than MC-LR or –LW. This would imply that the mOatps transporting ES also had a higher affinity for MC-LF than for the other MC congeners.

Finally, MC-LR, -LW, and -LF were demonstrated to have comparable PP-inhibitive capabilities (Monks et al., 2007; Fischer et al., in press) i.e. toxicodynamic properties. The fact that MC-LF was the MC congener that induced the greatest reduction in PP-activity following exposure of whole neuronal cells to equimolar concentrations of the three MC congeners, suggests that MC-LF was transported more efficiently into the cell (toxicokinetics) and thus reached PP-inhibitive intracellular concentrations more quickly than did MC-LR and –LW. More efficient transport of MC-LF was already reported previously in OATP transfected HeLa (Monks et al., 2007), HEK293 cells (Fischer et al., in press) and in murine whole brain cells containing a mixture of neuronal cells, astrocytes and microglia (Feurstein et al., 2009).

However, despite all of the indirect evidence presented above it has to be elucidated whether neuronal PP inhibition could result in neurotoxicity *in vivo*. Preliminary supportive *in vitro* data suggest a MC congener dependent degeneration of neurite networks at concentrations that did not induce overt cytotoxicity (Feurstein D., unpublished data).

Conclusion

All together the data presented here clearly revealed that mOatps play a crucial role in the neuronal uptake of MCs and that the uptake and subsequent neurotoxicity is high MC congener dependent. If these observations, including the higher potential toxicity of MC-LF, can be extrapolated to the human, they not only explain the observed neurotoxicity in the Caruaru accident, but also highlights the problems of using only one MC congener, MC-LR (WHO, 1999; Grosse *et al.*, 2006), as the basis for microcystin related human risk assessment.

Acknowledgements

Special thanks are due to Prof. Marcel Leist (University of Konstanz, Germany) for providing material and technical support and to Dr. Bernhard Ernst for critically reading the manuscript. We would like to acknowledge the International Research Training Group 1331 for financial support.

4. Manuscript III

MICROCYSTIN CONGENER AND CONCENTRATION DEPENDENT INDUCTION OF NEURITE DEGENERATION AND APOPTOSIS

Feurstein D.[1], Kleinteich J.[1], Stemmer K.[1], Speicher T.[2], Dietrich D.R.[1]

[1]*Human and Environmental Toxicology, University of Konstanz, Konstanz, Germany*
[2]*Biochemical Pharmacology, University of Konstanz, Konstanz, Germany*

Manuscript in preparation for publication in European Journal of Neuroscience

Abstract

Amongst other specific characteristic abnormal phosphorylated Tau protein and subsequent apoptosis are two hallmarks of Alzheimer's Disease (AD). Recent *in vitro* and *in vivo* studies have indicated that diarrhetic shellfish toxin okadaic acid (OA) triggers Tau hyperphosphorylation and caspase dependent cell death, mimicking the neurodegeneration observed in AD brain. Like OA, but with different cellular uptake mechanisms, the toxic cyanobacterial microcystins (MCs) act as strong serine/threonine-specific protein phosphatase (PP) inhibitors. Additionally, MCs have been suggested to be actively transported across the blood-brain-barrier and into neurons in a MC congener dependent manner thus acting as potent neurotoxins. However MCs have yet not been investigated whether they cause neurite degeneration and caspase dependent apoptosis as observed in AD. In this study cerebellar granule neurons (CGNs) were exposed to three individual MC congeners. A MC congener dependent abnormal phosphorylation of neuronal Tau epitopes at serine199/202 and serine396 was observed along with a significant decline of the neurite length (MC-LF >> MC-LW > MC-LR) already at low not cytotoxic concentrations. MC induced apoptosis, determined by nuclear morphology and caspase activity, was only significant for MC-LF and MC-LW at higher concentrations reducing neurite length by ~40%. Moreover, apoptosis was significantly reduced in the presence of the caspase inhibitor zVAD-fmk. These results demonstrate the neurotoxicity of MCs At equimolar concentration the MC congener MC-LF is the most potent variant. More subtle neurotoxicity, e.g. reduction of neurite length, develops already at concentrations which do not mediate apical responses, e.g. apoptosis, and thus may be an ideal parameter to study the subchronic neurotoxicity of MCs.

Keywords: Microcystin, cyanobacterial toxin, okadaic acid, neurotoxicity, Tau hyperphosphorylation, neurite degeneration, apoptosis, caspase

Introduction

In neurodegenerative diseases such as Alzheimer's Disease (AD) the microtubule (MT) stabilizing protein Tau is abnormal phosphorylated, thus leading a retrograde degeneration of the axo-dendritic network, decrease of the synaptic plasticity with subsequent induction of caspase dependent apoptosis (Cribbs *et al.*, 2004; K. Iqbal, 2008).

The cellular phosphorylation status was demonstrated to appear crucial for the induction and progression of tauopathies, confirmed by a reduction of PP1 and PP2A activities by approximately 20% in AD brain (Gong *et al.*, 1993; Gong *et al.*, 1995). Furthermore, Tau is phosphorylated at more than 30 serine/thronine (ser/thr) residues (Hanger *et al.*, 1998) by a dozen of kinases and is dephosporylated by at least four protein phosphatases (PP) (Gong *et al.*, 2005; Liu *et al.*, 2005). More recently, glycogen synthase kinase-3 (GSK-3β), c-jun N-terminal kinase (JNK) as well as PP2A and to a less extend PP1, were demonstrated to be key players in regulating phosphorylation and dephosphorylation of Tau (Liu *et al.*, 2005; Naruhiko Sahara, 2008).

One of the routinely used components to study the phosphorylation homeostasis in neuronal cells is okadaic acid (OA), a cell membrane permeable marine toxin which inhibits ser/thr-specific PPs. More recently it could be demonstrated that OA induced PP inhibition results in an *in vitro* and *in vivo* Tau hyperphosphorylation, caspase dependent cell death and subsequent memory loss, similar to that observed in AD (Rossini *et al.*, 2001; Pei *et al.*, 2003; Maidana *et al.*, 2006 2000; Yoon *et al.*, 2006).

Another group of selective ser/thr-specific PP inhibitors are the cyanobacterial microcystins (MCs) which require, unlike OA, an active uptake to enter cells via organic anion transporting polypeptides (rodent Oatps / human OATPs) (Fischer *et al.*, 2005; Komatsu *et al.*, 2007; Monks *et al.*, 2007). Indeed a MC congener dependent uptake into primary murine neurons was demonstrated more recently via murine Oatps (mOatps) (Feurstein *et al.*, 2009). Once MCs, of which more than 80 structural congeners have been identified (Meriluoto and Spoof, 2008) have crossed cell membranes, they mainly inhibit PP1 and PP2A by covalent binding to the catalytic PP-subunit (MacKintosh *et al.*, 1990; Hastie *et al.*, 2005). Consequently, the cellular phosphorylation balance become disrupted which resulted in hyperphosphorylation of numerous proteins (e.g. keratin 8 and 18), activation of kinases (e.g. Ca^{2+}/calmodulin-dependent protein kinase II), disruption of the cellular structure and subsequent cell death (apoptosis/necrosis) in a time and concentration dependent manner (Wickstrom *et al.*, 1995; Fladmark *et al.*, 1999; Toivola and Eriksson, 1999; Fladmark *et al.*, 2002; Batista *et al.*, 2003). More recent *in vitro* and *in vivo* studies have suggested MCs to possess memory loss and toxicity (Fischer *et al.*, 2005; Maidana *et al.*, 2006; Feurstein *et al.*, 2009). The latter suggestion is most likely also true for MC intoxicated humans, confirmed by immediate signs of neurotoxicity (e.g. tinnitus, mild deafness, visual disturbance) after acute MC exposure (Pouria *et al.*, 1998; Carmichael *et al.*, 2001; Azevedo *et al.*, 2002).

Based on the presence of Oatps/OATPs at the blood-brain-barrier and in neuronal membranes capable for transporting MCs as well as the similar mode of action compared to OA, it is thus likely that MCs exhibit neurodegeneration with similar characteristics than observed in AD. However, MCs have never been tested whether they cause neurite degeneration and caspase dependent apoptosis in neurons. Consequently, primary murine neurons were employed to determine cytotoxicity, apoptosis, neurite length and abnormal phosphorylated Tau in a MC congener and concentration dependent manner.

Materials and methods

Materials

Microcystin-LR, -LW and -LF were purchased from Alexis Biochemicals (Lausen, Switzerland). Cell culture media and reagents were purchased from PAA Laboratories (Pasching, Austria), culture plates from Greiner Bio-One (Frickenhausen, Germany), poly-L-lysin coated 8-well chamber slides from Becton Dickinson (Heidelberg, Germany) and cover slips from Thermo Scientific (Waltham, USA). All other chemicals and antibodies unless otherwise stated were from Sigma-Aldrich (Taufkirchen, Germany).

Isolation and neuronal cell culture

Balb/c mice were obtained from The Jackson Laboratory (Bar Harbor, U.S.A) and held at the animal facility (University of Konstanz, Germany) under standard conditions. Sacrifice and organ removal was carried out in accordance with the German Animal Protection Law (registry number: T-07 05). Primary murine cerebellar granule neurons (CGNs) were prepared from five to eight-day-old specific pathogen-free Balb/c mice as previously described (Feurstein *et al.*, 2009). Briefly, dissociated neurons were plated on either poly-L-lysine coated plates (50 mg/l) or culture slides at a density of about 2.0×10^6 cells/ml and cultured in BME supplemented with 10% FCS, 20 mM KCl and 1% penicillin-streptomycin (37°C, 5% CO_2). BME medium was renewed after 24 h containing 10 µM cytosine arabinoside. The isolated CGNs were cultivated three days in vitro (DIV3) and subsequently exposed to MCs, OA and Staurosporine (Stsp), without any further medium changes.

Cytotoxicity assay

Neurons were exposed to individual MC congeners for 48 h with varying concentrations and analyzed for cell viability using the standard 3-(4,5 dimethylthiazol-2-yl)-2,5-diphenyl tetrazolium bromide (MTT) reduction assay as previously described (Feurstein et al., 2009). TWEEN (0.1%) was applied as an internal positive control (0% viability) and untreated CGNs as a negative control (100% viability). Individual samples were analyzed in replicates (N=3 to 5), each with technical duplicates

Colorimetric PP inhibition assay
Neurons were either treated with 3 µM of single MC congeners for 24 to 56 h or with 6.7 nM OA for 4 to 32 h (Table 4.1) and subsequently analysed for total PP activity as previously described (Feurstein et al., 2009). Briefly, CGNs were homogenized in 80 µl ice cold enzyme solution buffer and equal protein amounts (1.7 µg) were transferred to a 96-well plate incubated with substrate solution and immediately measured at 405 nm (blank value). Plates were kept in a 37°C incubator for 90 min and measured again. Individual samples were analyzed in triplicates (N= 3) and the reduction of total PP activity in MC treated cells was calculated as percentage loss of PP activity compared with neurons (100% activity).

Assessment of nuclear morphology
For the determination of apoptosis, CGNs were exposed to varying concentrations of single MC-LR, -LW and -LF for 48 h. Two positive controls, OA and Stsp, were employed, treated for 24 h and subsequently monitored for nuclear morphology. Briefly, neurons were stained with 2.5 µM Hoechst 33342 for 10 min at RT and subsequently fixed with 2% PFA for 20 min at RT. Cover slips containing the stained cells were carefully transferred in a droplet fluorescent mounting medium upon microscope slide and subsequently observed for chromatin condensation as well as nuclear size and shape using fluorescence microscopy (Axiovert 200M, Zeiss, Göttingen, Germany). For caspase-3/7 inhibition experiments 15 µM of zVAD-fmk (Calbiochem, Merck Darmstadt, Germany) was employed 1 h before single MCs were added and subsequently investigated for nuclear morphology as described above. Individual samples were analyzed in triplicates (N= 3) and each replicate consists of 4 to 9 randomly taken pictures. That corresponds to approximately 250 cells per replicate. The number of apoptotic neurons was expressed as the percentage of total neurons, positive for Hoechst 33342.

Determination of caspase-3/7 activity
Neurons were treated with single MC congeners in a concentration (48 h) or time (3 µM) dependent manner und subsequently analyzed for caspase-3/7 activity. In addition, 6.7 nM of OA or 40 nM of Stsp were employed as internal positive controls (24 h). Cells were homogenized in 100 µl of ice cold caspase buffer A (25 mM HEPES, pH 7.5; 5 mM $MgCl_2$, 0.1 % (v/v) Triton X100) containing

(1 tablet/10 ml) complete protease inhibitor (Roche, Basel, Switzerland). Equal volumes (80 µl) of the homogenate were transferred to a 96-well plate and 80 µl caspase substrate buffer (50 mM HEPES, pH 7.4; 1 % (w/v) sucrose, 1 % (w/v) CHAPES, 50 µM Asp-Glu-Val-Asp-7-amino-4-trifluormethylcoumarin (DEVD-AFC), 10 mM DTT) was added. Release of AFC was measured fluoremetrically at 400 nm (Multilabelcounter, Wallac Victor2, Perkin Elmer, Rodgau-Jügesheim, Germany) over a time period of 1 h at 37°C. Specific activity was calculated by using an AFC standard curve. Values were finally normalized to the total protein content of the lysate (using Pierce Assay, Fischer Scientific, Schwerte, Germany) and the formation of 1 pmol AFC / min*mg protein was defined as 1 µU. Individual samples for MCs and OA/Stsp were analyzed in replicates (N= 3 to 6 and N= 4 to 9, respectively).

Western-blot (WB) analysis for abnormal phosphorylated Tau
Neurons were exposed to 3 µM of single MC congeners or 6.7 nM OA for varying exposure times, homogenized and visualized as previously described (Feurstein *et al.*, 2009). Briefly, protein samples (7 µg for MC and 3.5 µg for OA treated neurons) were separated by SDS-PAGE, blotted onto a nitrocellulose membrane and subsequently incubated with primary antibodies (AT8 1:1000 (Innogenetics, Gent, Belgium), PHF1 1:1000 (Millipore, Schwalbach, Germany) and GAPDH 1:30000) for 16 h at 4°C. Secondary antibodies were horseradish peroxidase conjugated goat anti-rabbit (1:160000) and rabbit anti-mouse (1:80000) incubated for 1 h at room temperature (RT). Immunoreactive bands were detected by enhanced chemiluminescence (GE Healthcare, Munich, Germany).

Investigation of neurite degeneration and image analysis
Neurons were treated with varying concentrations of single MC congeners for 48 h, fixed in 4% paraformaldehyd (PFA; 15 min, RT) and subsequently permeabilzed with PBS containing 0.2 % (v/v) Triton X100 for 10 min at RT. After blocking (PBS containing 1% BSA) for 1 h at RT, neurons were incubated with anti-class III β-tubulin (1:800) for 16 h at 4°C and subsequently incubated with fluorochrom-conjugated goat anti rabbit-ALX488 (1:1000) for 1 h at RT, both antibodies a kindly gift of Prof. Marcel Leist (University of Konstanz, Germany). Nuclei were counterstained with 2.5 µM Hoechst 33342 (Invitrogen, Karlsruhe, Germany) for 10 min at RT, mounted with fluorescent medium (DAKO, Hamburg, Germany) and subsequently visualized using a fluorescence microscope (Axiovert 200M, Zeiss, Göttingen, Germany). For image analysis and subsequent neurite length calculation an adapted version of the neurite tracer plugin for Image J provided by the Bioimaging Center of the University of Konstanz (http://bicwiki.kreeloo.de/index.php?title=ImageJ) was employed. Neurite length of MC treated

cells was compared to untreated cells for determining statistical significance. Individual samples were analyzed in triplicates (N= 3) and each replicate consists of 4 to 9 randomly taken pictures. That corresponds to approximately 300 cells per replicate.

Statistical analysis

Statistically significant differences were determined using Prism statistical software (GraphPad Prism 4.03, San Diego, U.S.A.). All data are presented as mean ± S.E.M of at least three independent experiments. For data analysis one- and two-way ANOVA with Dunnett's or Bonferroni's post test were applied. Statistical significance was set at P<0.05 (*) and P<0.01 (**).

Results

Investigation of MC congener and concentration dependent cytotoxicity

Neuronal cell viability was assessed after concentration dependent exposure to individual MC congeners for 48 h. A significant reduction of neuronal cell viability was already demonstrated at 0.8 µM MC-LF (81% viability) compared with control cells (100% viability) and was continuously reduced to 8% viability at the highest concentration (5 µM) (Figure 4.1). Moreover, neuronal exposure to 3 µM MC-LW revealed a significant reduction of cell viability by 63% which was further decreased to approximately 50% viability at 5 µM (Figure 4.1). The only significant induction of cytotoxicity after MC-LR exposure was observed at 5 µM, reducing neuronal viability to approximately 70% compared to untreated cells (Figure 4.1).

4. Manuscript III

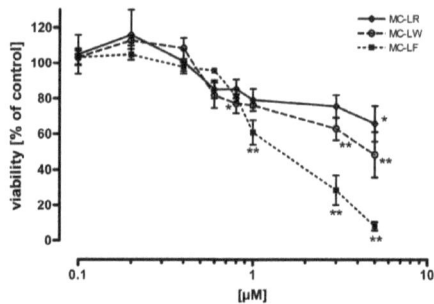

Figure 4.1:
Investigation of MC congener and concentration dependent cytotoxicity. Neurons were exposed to 0.1, 0.2, 0.4, 0.8, 1, 3 and 5 µM of single MC-LR, MC-LW and MC-LF for 48 h and analyzed for cell viability. Individual samples were analyzed in replicates (N= 3 to 5), each with technical duplicates and given as mean values ± SEM. Values are expressed as percentage of control (untreated neurons, 100% viability). One-way ANOVA with Dunnett's post-test; P<0.05 (*), P<0.01 (**).

Neuronal PP inhibition by MCs

A maximum and constant inhibition of total neuronal PP activity by approximately 50% was demonstrated for all three MC congeners as well as for the cell membrane permeable OA, between 24 to 56 h and 4 to 32 h, respectively (Table 4.1). The inhibitory effect of MC-LR, -LW and –LF was not significantly different from each other at all investigated time points.

Table 4.1: Total PP activity determination of MC-LR, -LW, -LF and OA treated primary neurons.

Toxin	PP activity (24 / 4* h) Mean ± SEM	PP activity (40 / 16* h) Mean ± SEM	PP activity (48 / 24* h) Mean ± SEM	PP activity (56 / 32* h) Mean ± SEM
MC-LR	47.0 ± 14.1	55.0 ± 8.1	49.5 ± 22.3	62.9 ± 13.9
MC-LW	42.8 ± 11.4	39.9 ± 10.7	53.0 ± 21.3	54.8 ± 19.0
MC-LF	57.1 ± 6.3	57.8 ± 8.5	78.1 ± 16.1	44.4 ± 20.1
OA*	42.1 ± 7.0	55.7 ± 14.2	67.9 ± 6.4	42.1 ± 11.6

Neurons were exposed to 3 µM of single MC congeners as well as to 6.7 nM OA and compared with untreated neurons (100% PP activity). Individual samples were analyzed in triplicates (N=3).

To further elucidate weather the PP inhibition of MC treated CGNs was reduced to a maximum level and to further exclude the membrane barrier of cultured CGNs, neuronal homogenates were employed, exposed to different MC-LR and –LF concentrations and revealed similar PP inhibition of approximately 40 - 50% for MC-LR and –LF, respectively (Table 4.2).

Table 4.2: Total PP activity determination of MC-LR and -LF treated homogenates of primary CGNs

Toxin	PP activity 0.3 µM Mean ± SEM	PP activity 0.6 µM Mean ± SEM	PP activity 1.3 µM Mean ± SEM	PP activity 2.5 µM Mean ± SEM
MC-LR	64.9 ± 9.0	66.0 ± 11.0	65.4 ± 9.7	73.6 ± 14.8
MC-LF	50.3 ± 2.0	47.3 ± 11.0	55.0 ± 9.0	51.5 ± 11.9

Neuronal homogenates were exposed to varying concentrations of single MC congeners for 30 min and compared with untreated neurons (100% PP activity). Individual samples were analyzed in triplicates (N=3), each with technical duplicates.

MC congener and concentration dependent induction of neuronal apoptosis

To characterize the apoptotic potential of single MC-LR, -LW and –LF in exposed CGNs (positive controls OA and Stsp), nuclear morphology (Figure 4.2 A and C) was investigated with/without the caspase inhibitor zVAD-fmk (Figure 4.2 B) as well as time and concentration dependent induction of caspase-3/7 activity (Figure 4.4 A, B and C).

Representative pictures of untreated (Figure 4.3 picture A) and 5 µM MC-LR, -LW and –LF treated neuronal nuclei (48 h) are shown in Figure 4.3 (pictures B, C and D, respectively) which demonstrated a MC congener dependent chromatin condensation at high concentrations.

Neuronal chromatin condensation was most prominent for MC-LF, with a significant increase in the percentage of apoptotic nuclei by 40% at 3 µM and by 60% at 5 µM (Figure 4.2 A). Neuronal cell death in MC-LW exposed CGNs was only significant at 5 µM with 32% apoptotic neurons (Figure 4.2 A). Interestingly, there was no apoptotic response from MC-LR treated neurons at any tested concentration (Figure 4.2 A) even at 10 µM MC-LR (data not shown). Confirmation of a MC induced programmed cell death was achieved by treating CGNs with 5 µM of single MC congeners with/without 15 µM zVAD-fmk (Figure 4.2 B). Neurons exposed to individual MC-LF and –LW together with zVAD-fmk, demonstrated a significant reduction of apoptotic cells to 27 and 17%, respectively (Figure 4.2 B). Apoptotic rates of MC-LF and –LW exposed CGNs together with the caspase inhibitor zVAD-fmk were slightly raised but not significantly different compared to untreated neurons (negative control). The highest apoptotic potential (90%) for all analyzed toxins (MC-LR, -LW, -LF, OA and Stsp) was observed for OA (20 nM, 24 h) treated CGNs (Figure 4.2 C). Additionally, Stsp exposed neurons revealed a maximum level at 40 nM (24h) with 55% apoptotic neurons (Figure 4.2 C) and did not differ up to 200nM (data not shown).

4. Manuscript III

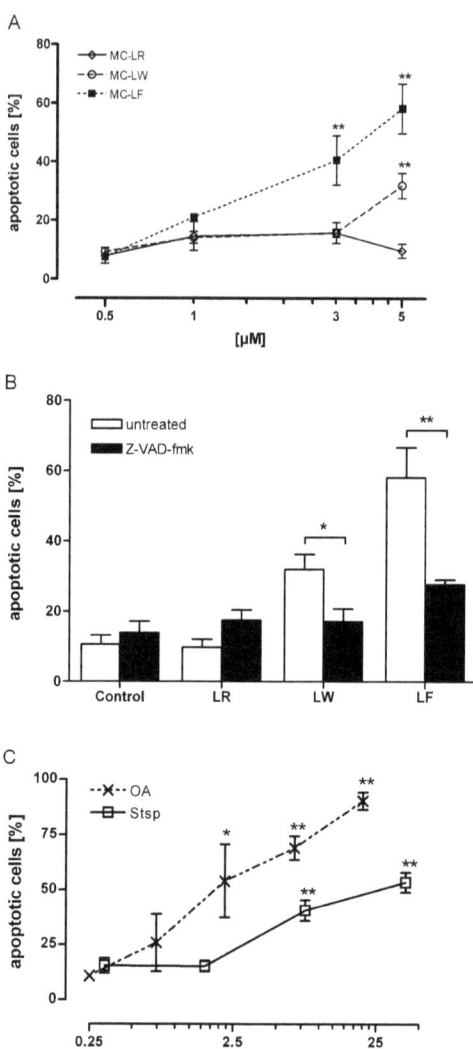

Figure 4.2:
Investigation of neuronal apoptosis determined by nuclear morphology. (A) CGNs were exposed to 0.5, 1, 3 and 5 μM of single MC-LR, -LW and –LF for 48h. One-way ANOVA with Dunnett's post test; P<0.01 (**). (B) Inhibition of MC induced apoptosis with the caspase inhibitor zVAD-fmk. CGNs were exposed to individual MC congeners (5 μM) with (black bars) and without (white bars) 15 μM of zVAD-fmk for 48 h. One-way ANOVA with Bonferroni's post test; P<0.05 (*), P<0.01 (**). (C) Neurons were treated with 0.4, 1.1, 2.2, 6.7, 20 nM OA and 0.3, 1.6, 8 and 40 nM Stsp for 24 h. One-way ANOVA with Dunnett's post test; P<0.05 (*), P<0.01 (**). (A-C) Individual samples were analyzed in triplicates (N= 3) and each replicate consists of 4 to 9 randomly taken pictures and represent mean values ± S.E.M.

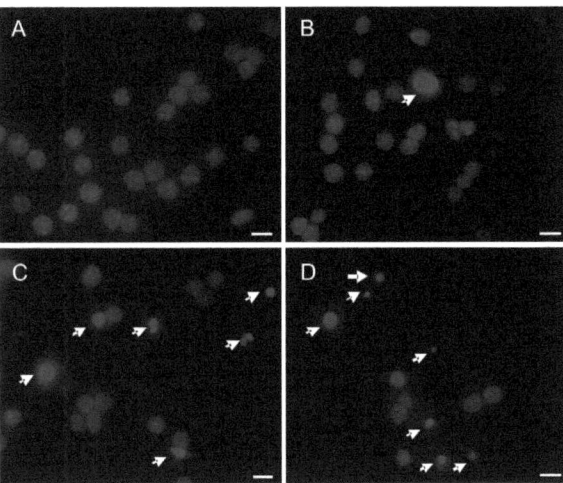

Figure 4.3:
Representative pictures of untreated and MC treated primary neurons. CGNs were exposed to 5 µM MC-LR (B), MC-LW (C) and MC-LF (D) for 48 h and subsequently stained for DNA (blue). Representative pictures of untreated neurons (A). Scale bar: 10 µm.

Apoptosis is per definition in its most classic form observed exclusively when caspases, in particular caspase-3, are activated (Leist and Jaattela, 2001). Indeed, caspase-3/7 activity was most prominent increased in MC-LF exposed CGNs with a significant dose dependent increase up to 300 µU at 3 µM and 800 µU at 5 µM, compared with untreated cells (~50 µU) (Figure 4.4 A). In addition, caspase-3/7 activation was approximately half as intense in MC-LW- compared to MC-LF- exposed CGNs, revealing a significant increase of caspase-3/7 activity (470 µU) only at 5 µM MC-LW (Figure 4.4 A). There was no effect observed for MC-LR treated neurons up to a concentration of 5 µM (Figure 4.4 A) whereas exposure to 10 µM MC-LR indicated a slight but significant induction of caspase-3/7 activity approximately 2-fold above control levels (data not shown). Caspase-3/7 activity was additionally determined in a MC congener (3µM) and time dependent manner and compared with the representing control neurons (Figure 4.4 B). The first significant effect was measured 44 h after MC-LF exposure (274 µU) with a maximal caspase-3/7 activity of approximately 776 µU after 52 h (Figure 4.4 B). In addition, MC-LW induced caspase-3/7 activity was significantly increased to 392 µU after 52 h with a maximal activity of 500 µU after 56 h (Figure 4.4 B). Contrary, neuronal exposure of 3 µM MC-LR did not induce caspase-3/7 activity at any tested time point (Figure 4B). The two positive controls OA (6.7 nM, 24 h) and Stsp

4. Manuscript III

(40 nM, 24 h) demonstrated a significant increase of the caspase 3/7 activity of more than 1000 and 800 µU, respectively (Figure 4.4 C).

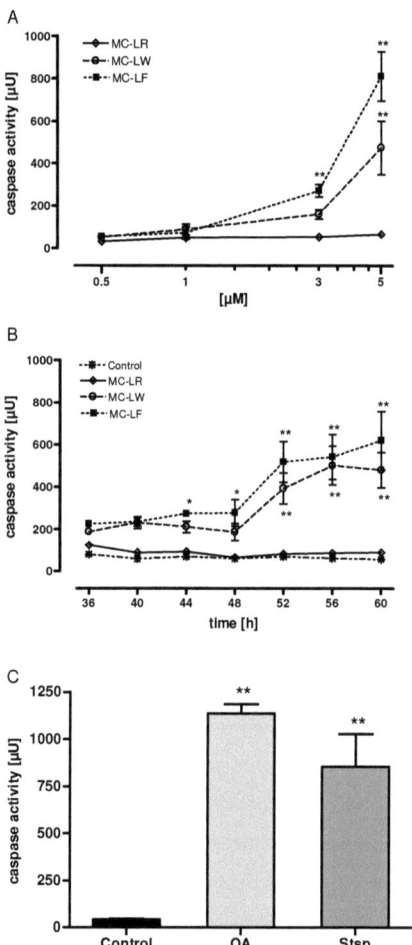

Figure 4.4:
Investigation of neuronal caspase-3/7 activity. (A) CGNs were exposed to 0.5, 1, 3 and 5 µM of single MC-LR, -LW and -LF for 48 h. Individual samples were analyzed in replicates (N=3 to 6) and given as mean values ± S.E.M. One-way ANOVA with Dunnett's post test; P<0.01 (**). (B) Neurons were exposed to 3 µM of single MC congeners for 36 to 60 h. Individual samples were analyzed in triplicates (N= 3), and given as mean values ± S.E.M. Two-way ANOVA with Bonferroni's post test; P<0.05 (*), P<0.01 (**). (C) CGNs were treated with 6.7 nM OA and 40 nM Stsp for 24 h. Individual samples were analyzed in replicates (N= 4 to 9), and given as mean values ± S.E.M. One-way ANOVA with Dunnett's post test; P<0.01 (**). (A-C) The formation of 1 pmol AFC / min*mg protein was defined as 1 µU.

4. Manuscript III

Determination of abnormal phosphorylated Tau after MC exposure

Neuronal exposure to 3 µM of single MC-LR, MC-LW and MC-LF for 24 and 40 h with subsequent WB analyzes revealed two immunopositive bands with 54 and 58 kDa for two tested Tau epitopes (Ser199/202, detected with AT8 antibody and Ser396, detected with PHF1 antibody) (Figure 4.5 A and B). Untreated CGNs were negative, except a light immunopositve band with 54 kDa at Tau-Ser396 (Figure 4.5 A and B). These results indicate that Tau became abnormal phosphorylated at MC concentrations reducing total PP activity by approximately 50% (Table 4.1). Treatment with 3 µM of single MC congeners for 48 and 56 h with subsequent WB analysis demonstrated weakened immunopositive signals (Figure 4.5 C and D) but most prominent only for MC-LF and to a lesser extend for MC-LW. Determination of OA-treated CGNs (positive control) revealed immunopositive bands with 54 and 58 kDa at Tau-ser199/202 and Tau-ser396, however only 4 h after exposure (Figure 4.5 A). It is suggested that the MC congener and time dependent reduction of the immunopositive signals was due to the strong cytotoxicity of MC-LF and –LW at 3 µM (Figure 4.1), most likely also true for OA.

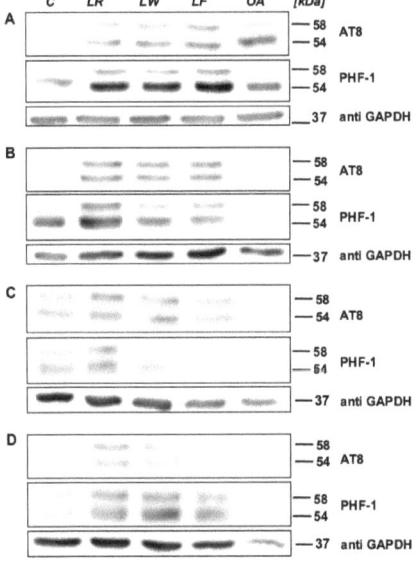

Figure 4.5:
WB analysis of abnormal phosphorylated Tau. Primary neurons were exposed to 3 µM MC-LR, MC-LW and MC-LF for (A) 24 h, (B) 40 h, (C) 48 h and (D) 56 h and with 6.7 nM OA for (A) 4 h, (B) 16 h, (C) 24 h, (D) 36 h. Phospho-specific Tau-antibodies AT8 (Ser199/202) and PHF1 (Ser396) and GAPDH as housekeeping protein were employed. "C" represents untreated CGNs (negative control).

MC congener dependent neurite degeneration

Untreated CGNs (negative control) were characterized by a dense microtubule network (Figure 4.7 pictures A and E) with an average neurite length of 144.5 µm per neuron. However, the neuronal network in MC treated CGNs became morphologically less distinctive in a concentration and MC congener dependent manner although PPs were similar inhibited already 24h after exposure (Table 4.1). Representative pictures of 1, 5 and 10 µM MC-LR treated neurons (Figure 4.7 pictures F to H) indicated only a slight reduction of the neurite network at highest concentration inducing moderate cytotoxicity (Figure 4.1) but no apoptosis (Figure 4.2 A and Figure 4.4 A). However, representative pictures of 1, 5 and 10 µM MC-LF exposed CGNs (Figure 4.7 pictures B to D) revealed a dramatic decline of the neurite network already at low concentrations (Figure 4.7 picture B) with moderate cytotoxicity (Figure 4.1) and initial apoptosis (Figure 4.2 A). At 5 µM, MC-LF exposed CGNs (Figure 4.7 picture C) exhibited only occasional neurites, not surprising in conjunction with the high apoptotic and cytotoxic potential (Figure 4.1, Figure 4.2 A and Figure 4.4 A) at this concentration.

Image analysis of MC treated neurons clearly demonstrated a concentration and MC congener dependent reduction of the neurite length 48 h after exposure (Figure 4.6). For MC-LR exposed neurons, a concentration of 3 µM was necessary to slightly but significantly reduce the neurite length by 17% whereas highest MC-LR concentrations (>8 µM) were necessary to reduce the total neurite length by 30% compared with untreated CGNs (Figure 4.6). Image analyses of MC-LW treated neurons demonstrated a significant reduction of the neurite length already at 1 µM (29%) whereas at the highest concentration (10 µM) neurite length was reduced by 70% compared with untreated CGNs (Figure 4.6). Based on the high cytotoxic and apoptotic potential of MC-LW already at 5 µM (Figure 4.1, Figure 4.2 A and Figure 4.4 A) it is suggested that the reduction of neurite length by 70% at 10 µM represents late apoptotic neurons. Interestingly, comparing the total neurite length of untreated neurons (144.5 µm) with MC-LF exposed neurons, already the lowest concentration (0.5 µM) used in this assay was sufficient to significantly reduce the neurite length per cell to 106.6 µm (26%) (Figure 4.6). At the same concentration MC-LF treated CGNs demonstrated no cytotoxicity (100% viability) and no induction of apoptosis (Figure 4.1 and Figure 4.2 A). At ≥8 µM MC-LF only neurite fragments were observed, which was not surprising in conjunction with a reduction of cell viability by 92% already at 5 µM.

4. Manuscript III

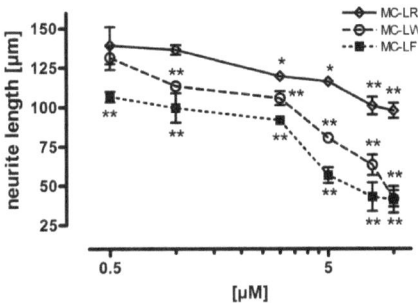

Figure 4.6:
Investigation of the neurite length after exposure to single MC congeners. Image analysis and neurite length determination of 0.5, 1, 3, 5, 8 and 10 µM MC-LR, MC-LW and MC-LF exposed CGNs (48h). Individual samples were analyzed in triplicates (N=3) and each replicate consists of 4 to 6 taken pictures and given as mean values ± S.E.M. One-way ANOVA with Dunnett's post test; $P<0.05$ (*), $P<0.01$ (**).

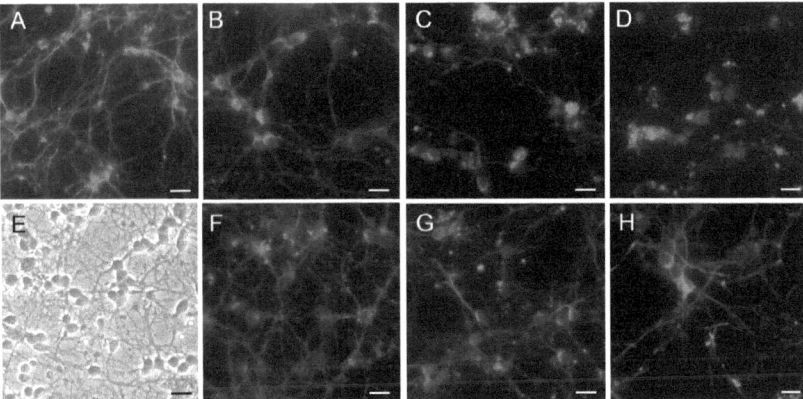

Figure 4.7:
Neurite length determination by immunocytochemistry. Representative pictures of untreated neurons (A and E), 1, 5 and 10 µM MC-LF (B, C and D, respectively) and MC-LR (F, G, and H, respectively) exposed CGNs. Primary neurons were stained for β-tubulin (green) and DNA (blue). Scale bar: 10 µm.

Discussion

The signal transduction hypothesis of neurofibrillary degeneration in tauopathies, including AD was published more recently by Iqbal and Grundke-Iqbal (K. Iqbal, 2008) and suggest that a variety of environmental factors are triggering a genetic predisposition, inducing a protein phosphorylation/dephosphorylation imbalance and subsequent abnormal hyperphosphorylation of Tau. Consequently, the microtubule structure becomes disrupted, thus revealing loss of neuronal function and the induction of caspase dependent apoptosis (K. Iqbal, 2008). This hypothesis raises the question weather a genetic predisposition is fundamental for the induction of tauopathies since the imbalance of phosphorylation could also be simply due to e.g. PP inhibition by MC or OA exposure.

However, unlike OA, MC induced neurotoxicity can only evolve if transporters are expressed at the BBB and additionally in neuronal cell membranes and capable for transporting MCs. So far, human OATP1A2 (expressed e.g. at the BBB), 1B1 and 1B3, rat and murine Oatp1b2 as well as scate Oatp1d1 were demonstrated to transport MCs across membranes (Fischer et al., 2005; Komatsu et al., 2007; Monks et al., 2007; Lu et al., 2008). More recently, mOatps dependent uptake of MC-LR, -LW and –LF into primary murine CGNs was indirectly confirmed and suggested strong variable uptake kinetics depending on the MC congener (see manuscript II).

MCs have only been suggested by several authors to represent possible neurotoxins but the latter was never proven directly in a neuron-specific *in vitro* model. In addition, the ser/thr specific PP inhibitor OA was demonstrated *in vitro* and *in vivo* to induce an Alzheimer's-like pathology. Based on the similar mode of action three variants of the cyanobacterial toxin MC were characterized in terms of their capability to induce neurite degeneration and caspase dependent apoptosis.

Indeed, we have clearly demonstrated a MC congener and concentration dependent induction of neurotoxicity which was most prominent for MC-LF (48h) (Figure 4.1). However, neuronal PP activity was equal inhibited by MC-LR, -LW and –LF already 24 h after MC exposure (Table 4.1). It is suggested that the significant reduction of the total PP activity in MC exposed CGNs has reached the maximum effect at ~50% of total PP activity. The latter was confirmed by treating CGNs with the cell membrane permeable toxin OA (Table 4.1) and CGN homogenates with single MC congeners (Table 4.2). Indeed neuronal PPs were inhibited to the same extend than MC exposed cultured neurons. Moreover, MCs are potent and selective inhibitors of ser/thr-specific PPs with comparable and low nM $IC_{50}s$ for PP1 and PP2A, especially true for MC-LR, -LW and –LF (Monks et al., 2007). Consequently, the observed total neuronal PP rest activity is most likely governed by other ser/thr-specific PP members which are to a lesser extend inhibited by MCs (e.g. PP2B and PP7) (Hastie et al., 2005) and by non-ser/thr-specific PPs. More recently, OA induced ser/thr specific PP inhibition was reported to target essential steps in the signaling machinery that governs the life and death of a cell (Clorinda Arias, 1993; Kim et al., 1999; Pei et al., 2003; Maidana et al., 2006). Moreover, MC-LR has also been reported to induce apoptosis *in vitro* and in *vivo*, but the exact mechanisms of MC-LR induced cell death are still unknown (Fu et al., 2005). Furthermore, not all tested cell-types undergo MC-LR induced apoptosis (Teneva et al., 2005).

However, the data presented here clearly demonstrate a MC induced caspase-3/7 dependent apoptosis at high MC concentrations, which was strongly dependent on the MC congener. While MC-LR was not able to induce apoptosis in primary CGNs at any tested concentrations, with the exception of a light induction of the caspase-3/7 activity at 10 µM, MC-LW and –LF followed a time and concentrations dependent cell death mediated by caspase-3/7 (Figure 4.2 A, Figure 4.4 A and B). Of all three tested MC congeners, MC-LF demonstrated the strongest apoptotic potential in primary neurons already 44 h after exposure (Figure 4.4 B).

The above discussed observations, including MC and OA induced PP inhibition and caspase-3/7 dependent cell death was confirmed by more recent investigations. It could be demonstrated that OA induced activation of JNK has triggered the mitochondrial death pathway via bim (Bcl-2 interacting mediator of cell death) over expression, caspase-3 activation (Yoon *et al.*, 2006) an additionally induced abnormal Tau hyperphosphorylation (Kim *et al.*, 1999) at the same Tau epitopes (Ser199/202 and Ser396) than observed in MC treated neurons (Figure 4.5). In addition, hyperphosphorylated Tau has been reported recently to be cleaved by caspase-3/7 *in vitro* and *in vivo*, and it was suggested by the authors that this event is representing an early event in AD tangle pathology (Guo *et al.*, 2004; Rametti *et al.*, 2004; Naruhiko Sahara, 2008). In MC-LR exposed HEK293 cells, stably expressing the human MC transporters OATP1B1 and OATP1B3, upregulation of active JNK and cytotoxicity was observed whereas its inhibition resulted in an increased of cell viability (Komatsu *et al.*, 2007). It can be concluded that inhibition of ser/thr specific PPs trigger neuronal cell death and it is suggested that JNK could also be involved in the observed neuronal apoptosis induced by MC-LF and MC-LW.

As PP2A and to a less extend PP1 are known to regulate the microtubule stabilizing Tau protein (Liu *et al.*, 2005), it is thus not surprising that the observed PP inhibition after MC and OA exposure resulted in Tau hyperphosphorylation (Figure 4.5) at two tested pathological epitopes (Ser199/202 and Ser396) with consequent neurite degeneration (Figure 4.6) in a MC congener and concentration dependent manner (MC-LF > -LW > -LR). Indeed, a degeneration of the neurite network could be demonstrated for MC-LF and MC-LW at non-cytotoxic concentrations, whereas MC-LR revealed a significant reduction of the neurite length together with a reduction of cell viability by 30% (Figure 4.6 and Figure 4.1). Thus, these results indicate different toxicokinetics for the three tested MC congeners, especially in conjunction with previous reports demonstrating varying uptake capabilities for MC-LR, -LW and –LF by mOatps (Feurstein *et al.*, 2009; see manuscript II). However, MC-LR was reported to alter all three cytoskeletal components, including microfilaments and intermediate filaments in non-neuronal cells (Wickstrom *et al.*, 1996; Toivola and Eriksson, 1999; Batista *et al.*, 2003). Therefore it is suggested that not only the demonstrated microtubule breakdown in MC exposed CGNs was leading to the observed neurite degeneration rather alter various cytoskeletal-associated proteins, like keratin 8/18.

All together, the data presented here clearly indicated that MCs are potent neurotoxins with MC-LF being the most toxic variant. Moreover, it is suggest that the observed MC congener differences in terms of neurite degeneration, cytotoxicity, and apoptosis are mainly dependent on the uptake capacity of functionally expressed mOatps in neuronal membranes with varying affinities for individual MC congeners. In conclusion, the present *in vitro* findings indicate that MCs are able to induce an Alzheimer's-like pathology at high concentrations. However, low MC-LF and MC-LW concentrations only revealed a degeneration of neurites and therefore represents an ideal parameter to study the subchronic neurotoxicity of MCs.

Acknowledgements

We would like to acknowledge the International Research Training Group 1331 (IRT1331) between Konstanz, Germany and Zurich, Switzerland for kindly funding this work. We would also like to thank Prof. Dr. Marcel Leist (Doerenkamp Zbinden Chair of Alternative *in vitro* Methods, University of Konstanz, Germany) for providing material and constructive discussions, Dr. Elisa May and Felix Schöneberger (Bio Imaging Centre, University of Konstanz, Germany) for the great support during image analysis as well as Dr. Bernhard Ernst for reading the manuscript.

5. Additional data

5.1. MC in vivo study

This chapter demonstrates additional data from acute MC treated Balb/c mice. These preliminary data were not presented in one of the previous manuscripts.

Investigation of acute neurotoxic effects of MC-LR and MC-LF in mice

The MC *in vivo* study was performed in collaboration with Dr. Valerie Fessard and Dr. Ludovic Le Hegarat, in the laboratories of the Agence Française de Sécurité Sanitaire des Aliments (AFSSA) – The French Food Safety Agency, Javené, France.

Hypotheses

- Based on the presence of Oatps at the blood-brain-barrier (BBB) / blood-cerebrospinalfluid-barrier (BCSFB) it was hypothesized that MCs are transported across the BBB/BCSFB in an Oatp/OATP as well as MC-congener dependent manner and will exert toxic effects in the brain, especially in neurons.

These hypotheses are supported by *in vitro* studies (Chapter 2, 3 and 4) which demonstrated the presence of various Oatps in primary mWBCs and mCGNs. These findings are corroborated by mouse brain *ex vivo* analyses of Cheng et al. (Cheng *et al.*, 2005). The concentration-dependent uptake and neurotoxic potential of three single MC-congeners (MC-LR/-LW/-LF) was shown in mWBCs and mCGNs (Chapter 2 and 4).

- Based on the *in vitro* studies, it was hypothesized that MC-LF is more toxic *in vivo* –than the well characterized MC congener MC-LR.

Indeed, the more hydrophobic variant MC-LF was shown to be the most toxic congener in mWBCs and mCGNs followed by -LW and –LR (Chapter 2 and 4). The latter was also confirmed with OATP1B1 and OATP1B3 transfected HEK293 (Fischer et al., in press).

- Furthermore it was hypothesized that once MCs have crossed the BBB/BCFB, MCs remain in the brain, covalently bound to PPs.

5. Additional data

So far MC or MC-conjugate export by MRPs or MDRs have not been conclusively demonstrated. The latter export however is assumed to be negligible in comparison to the actual concentration of MCs present in any given tissue as covalent binding of MCs to sulfhydril-groups of Cys273/266 of PPs (catalytic subunit) would compete with MC-metabolism/conjugation and export. Consequently, subchronic to chronic administration of MCs should result in a progradient accumulation of MCs in the brain as well as other organs of exposed animals and only the rate of cell regeneration in conjunction with cell debris phagocytosis and/or excretion will influence the rate of MC accumulation in a given organ.

In order to test the hypotheses outlined above, the goal of this study was to determine the accumulation, localization and the neurotoxic effects of MC-LR and MC-LF in male and female Balb/c mice.

Experimental setup

Single MC congeners (MC-LR and MC-LF) used in this experiment were of the highest analytical grade commercially available and obtained from Alexis Biochemicals (Lausen, Switzerland).

Six to eight week old Balb/c mice of approximately 25-30 g were chosen because the more recent MC *in vivo* studies have used Balb/c mice (Ito *et al.*, 2000 2004) which will yield a better comparison between *in vivo* and *in vitro* (mWBCs and mCGNs). In addition, it was shown that Balb/c mice are more sensitive than ICR mice to MC-LR (Ito *et al.*, 2000). Male and female mice were chosen because of the gender dependent differences in the Oatp distribution and organ specific expression (Cheng *et al.*, 2005).

In order to reach steady-state conditions, mice were i.p. injected with 20 µg MC-LR or MC-LF /kg bw every other day. The final concentration of 140 µg MC-LR or MC-LF /kg bw was estimated from published peer-reviewed chronic i.p. and p.o. *in vivo* studies using MC-LR (Solter *et al.*, 1998; Fawell *et al.*, 1999). Table 5.1 summarizes the experimental setup.

5. Additional data

Table 5.1: Experimental setup

group 1 number of animals	toxin	every second day	total concentration	application	weeks
10 male	MC-LR	20µg/kg	140µg/kg	i.p.	2
10 female	MC-LR	20µg/kg	140µg/kg	i.p.	2
10 male	MC-LF	20µg/kg	140µg/kg	i.p.	2
10 female	MC-LF	20µg/kg	140µg/kg	i.p.	2
10 male	saline			i.p.	2
10 female	saline			i.p.	2
group 2 number of animals	toxin	every second day	total concentration	application	weeks
10 male	MC-LR	20µg/kg	140µg/kg	i.p.	2+2 pe
10 female	MC-LR	20µg/kg	140µg/kg	i.p.	2+2 pe
10 male	MC-LF	20µg/kg	140µg/kg	i.p.	2+2 pe
10 female	MC-LF	20µg/kg	140µg/kg	i.p.	2+2 pe
10 male	saline			i.p.	2+2 pe
10 female	saline			i.p.	2+2 pe
positive control number of animals	toxin	single injection	total concentration	application	weeks
2 male	MC-LR	100µg/kg	100µg/kg	i.p.	0
2 female	MC-LR	100µg/kg	100µg/kg	i.p.	0
2 male	MC-LF	100µg/kg	100µg/kg	i.p.	0
2 female	MC-LF	100µg/kg	100µg/kg	i.p.	0

pe: post exposure

Organs (brain, liver and kidneys) were removed with the help of animal facility members of AFSSA and Prof. Dr. Daniel R. Dietrich and either fixed in paraffin for immunohistochemistry, in liquid nitrogen for protein analysis or in RNA stabilizing solution (RNAlater, Quiagen, Hilden, Germany) for further RNA analysis. For protein and RNA analysis, the brain was separated into cerebellum, cortex and "rest". The whole blood was collected and separated to serum/plasma fractions at the end of the experiment and stored in liquid nitrogen.

Methods

Detection of MCs by Western-blot (WB) analysis

MC WB analysis was carried out as previously described (Chapter 2) with the only difference that brain fragments were homogenized in 1200 µl extraction buffer. Monoclonal antibody to MCs (Adda-specific, clone AD4G2 from Alexis Biochemicals, Lausen, Switzerland) was diluted 1:1500 in blocking buffer.

Detection of MCs by immunohistochemistry (IHC)

Brain and liver slides (kindly provided from AFSSA) were deparaffinised in 100% xylene, and rehydrated in decreasing concentrations of ethanol (100%, 95% and 70%). Endogenous peroxidase was blocked with 3% H_2O_2 for 10 min at RT. Endogenous biotin was blocked for 15 min at RT using a commercial available avidin/biotin blocking reagent (BioGenex, San Ramon, USA) and additionally incubated with Powerblock (BioGenex, San Ramon, USA) for10 min at RT. The primary antibody sheep anti-Adda (Fischer *et al.*, 2001) was diluted 1:5000 in Powerblock and subsequently incubated with samples for 1 h at RT in a humidified atmosphere. Thereafter, biotinylated secondary antibodies (BioGenex, San Ramon, USA) were incubated with samples for 30 min at RT in a humidified atmosphere and the antigen- antibody complexes were visualized using a HRP-labeled, biotin-streptavidin amplified detection system (BioGenex, San Ramon, USA) with AEC (3-amino-9-ethylcarbazole) chromogen as a substrate (7 min at RT). Slides were counterstained with Mayer's hematoxylin (Sigma-Aldrich, Taufkirchen, Germany) for 15 min at RT, rinsed with tap water and mounted using CrystalMount (Biomedia, Manistique, USA) and Shandon Histomount (Thermo Scientific, Braunschweig, Germany).

5.2. Preliminary results and discussion

The above described immunohistochemical method indicated the best results in terms of no background staining and intense MC-immunopositive signals (Figure 5.1) among different tested staining methods. However, even the method described above, did not lead to constant immunopositive staining in slices (brain and liver as a positive control) of MC-LR or MC-LF exposed animals and additionally not always indicated negative results in brain and liver slices of untreated mice. It was suggested that the quality of the sections as well as the deparaffination of tissue slides seem to be crucial for a specific staining. Due to the lack of analyzed replicates and to varying results it can only be suggested that MCs do cross the BBB and accumulate in the brain (Figure 5.1 B tand C). In comparison to immunopositive brain sections, the liver demonstrated an intense staining in the cytosol of hepatocytes for both MC-LR (Figure 5.1 E to F) and MC-LF (Figure 5.1 H and I) exposed mice (140 µg MC/kg over two weeks).

Based on the preliminary IHC data presented here and more recent MC organ distribution studies (Meriluoto *et al.*, 1990; Hall *et al.*, 2008) it can be concluded that MCs are most likely able to enter brain and accumulate in different brain regions.

5. Additional data

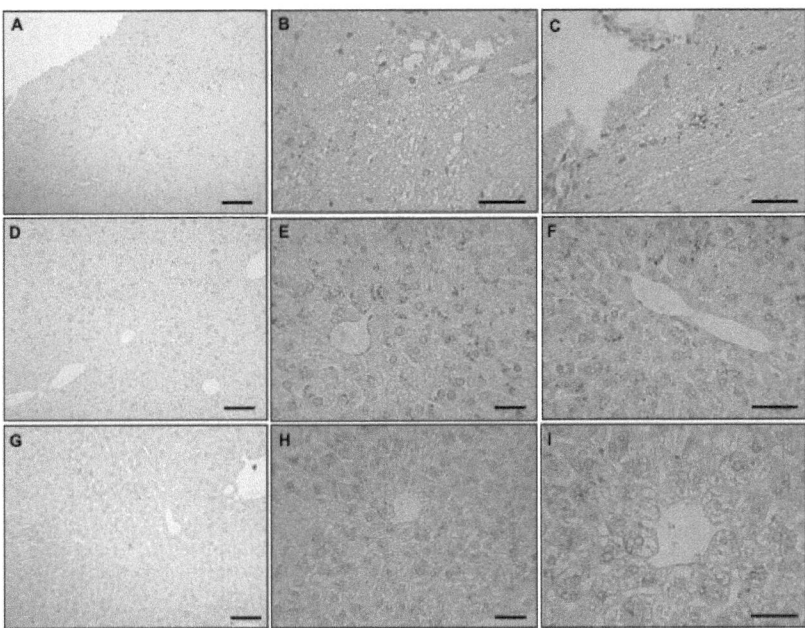

Figure 5.1:
Immunohistochemical detection of MCs in brain and liver paraffin sections of untreated, MC-LR and MC-LF exposed male mice (acute study). (A) Brain section of untreated and (B and C) MC-LR treated mice. (D and G) Liver sections of untreated mice. (E and F) Liver sections of MC-LR and (H and I) MC-LF exposed mice. Scale bar: 100 µm.

For MC-LR and MC-LF detection in brain and liver homogenates, different primary antibodies were employed and tested for specific binding. Among all tested antibodies Adda-specific clone AD4G2 was finally selected and demonstrated MC immunopositive bands in brain (38.3 kDa; lane 1-3) and liver (33 kDa; lane 4) homogenates of the same untreated animal (Figure 5.2; lanes 1-4). Homogenates of the brain cortex (lane 5) of MC-LF exposed mice revealed the same MC immunopositve band that was observed for the untreated mice and at 36.7 kDa a signal in the liver homogenate (lane 6). *In vitro* incubated MC-LF-PP1α (lane 7) and –PP2A/C (lane 8) adducts were employed as controls to compare the molecular weights with adducts observed in the MC *in vivo* study (Figure 5.2; lane 7 to 8). The MC-LF-PP1α and –PP2A/C adducts revealed immunopositive bands at approximately 37 kDa and therefore fit only with the molecular weight (liver homogenate, lane 6) of MC-LF exposed mice.

5. Additional data

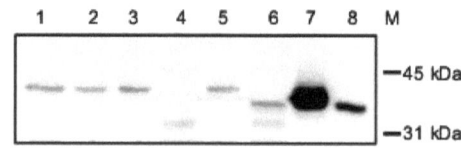

Figure 5.2:
Immunodetection of MC-LF in brain and liver homogenates of exposed and untreated male mice. Lane 1-4: untreated mice representing brain cortex, cerebellum, rest and liver samples, respectively. Lane 5 and 6: MC-LF exposed mice representing cortex and liver samples, respectively. Lane 7 and 8: recombinant PP1α and PP2A incubated with MC-LF for 30 min (positive control).

In addition to the above described preliminary results of the MC *in vivo* study, internal negative and positive controls of the brain and liver homogenates were tested (Figure 5.3). The mice used for this experiment were obtained from the animal facility (University of Konstanz, Germany) and organ removal was carried out in accordance with the animal protection law (registry number: T-07 05). Brain homogenates (internal negative control) revealed no signal (lane 1). MC-LR and MC-LF incubated brain homogenates indicated a MC immunopositive band at 36.7 kDa (lane 2 and 3, respectively). The latter was also observed for MC-LR and MC-LF incubated liver homogenates (lane 5 and 6), however with a slightly reduced molecular weight (~35 kDa).

Based on the demonstrated preliminary results (Figure 5.2 and Figure 5.1) it can be concluded that MC-PP1α and –PP2A adducts demonstrate molecular weights of approximately 35-37 kDa. Indeed, PP1α and PP2A/C have molecular weights of 37.5 and 36 kDa, respectively, yet in addition containing a covalently bound 1 kDa MC. However, untreated brain and liver homogenates (Figure 5.2; lane 1 to 4) of the *in vivo* study should be negative as demonstrated below in internal negative samples (Figure 5.3; lane 1 and 4). More recently, an unspecific binding of the AD4G2 clone (Adda-specific) was suggested (personal communication with Dr. V. Fessard) which could explain the discrepancy in the above presented preliminary results (Figure 5.2). However the specific binding of the Adda-specific antibody demonstrated below (Figure 5.3) would exclude this suggestion.

For further IHC and WB analysis of the *in vivo* samples it seems to be crucial to test other MC-specific antibodies (which will be provided by AFSSA, France) and to confirm MC immunopositve signals with analytical methods.

5. Additional data

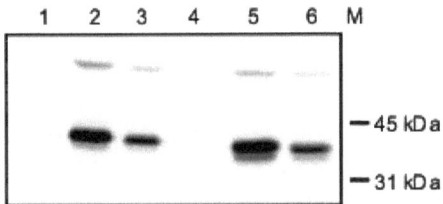

Figure 5.3:
Internal negative and positive controls. Brain and liver homogenates of untreated mice were incubated with and without 1 µM MC-LR or MC-LF for 30 min. Lanes 1 to 3: Untreated, MC-LR and MC-LF exposed brain homogenates, respectively. Lane 4 to 6: Untreated, MC-LR and MC-LF exposed liver homogenates, respectively.

6. Overall discussion

6.1. Toxic cyanobacteria: a growing risk?

Cyanobacteria are abundant in marine, brackish and freshwaters, including rivers, lakes, ponds and drinking water reservoirs (Chorus and Bartram, 1999; Briand et al., 2003). During the past 131 years, dating back to George Francis's first publication in 1878 (Francis, 1878), numerous cyanobacterial poisonings of aquatic and terrestrial animals as well as humans (Kuiper-Goodman et al., 1999; Carmichael, 2008; Stewart et al., 2008) had been reported worldwide. This is especially true for MCs, in conjunction with contaminated drinking water and food, including blue-green algae-based dietary supplements (BGAS) and recreational activities, like sauna, swimming, boating and water skiing in cyanobacteria blooming water (Dietrich et al., 2008).

Indeed, based on reported blooming events and published scientific investigations, their incidence has exponentially increased since the 1970's (Carmichael, 2008). The observed phenomenon could be explained by three major aspects. First of all, current research suggests that climate change, specifically global warming, will enhance cyanobacterial mass occurrences (Paerl and Huisman, 2008; Paul, 2008). Furthermore, the relationship between the worldwide increasing eutrophication of water bodies and the severity of the global occurrence of MC producing species, mainly *M. aeruginosa*, is well accepted (de Figueiredo et al., 2004; Carmichael, 2008). Finally, artificial water reservoirs (e.g. Itaparica Dam, Brazil and Nhlanganzwane Dam, Kruger National Park, South Africa) seem to encourage *Microcystis* mass profiferation (Teixera et al., 1993; Oberholster et al., 2009).

Consequently, it has to be assumed that the latter aspects would also indicate an increasing health hazard to humans by drinking water and taking part in recreational outdoor activities. In addition, more frequent toxin accumulation and possibly higher amounts of MCs in food sources (e.g. fish, prawns), including BGAS, is likely. Therefore, it can be concluded that toxic cyanobacteria, especially true for MC producing species, represent a growing risk for human health and livestock.

6.2. Microcystins – potent neurotoxins?

Initial situation

Many scientists in the field call MCs "hepatotoxins". This statement is based on the primary distribution and toxicity of MCs targeted mainly the liver (Hooser et al., 1989; Weng et al., 2007) but also, even to a lesser extent, the intestine, kidney, spleen and muscles (Hooser et al., 1989; Meriluoto et al., 1990; Nishiwaki et al., 1994; Botha et al., 2004; Hall et al., 2008).

6. Overall discussion

More recently and not surprising, initial evidence emerged suggesting MCs as possible neurotoxins (Chapter 1.6) likely true for mammals, including humans. However, precise information about accumulated amounts of MC-LR in the brain of laboratory mice is limited due to the following reasons. All three listed MC-LR distribution studies (Table 1.4) were carried out using ^{3}H-dihydro-MC-LR, a variant demonstrated to be less toxic compared to MC-LR (Robinson *et al.*, 1989; Namikoshi *et al.*, 1993; Zurawell *et al.*, 2005). Furthermore, two different epimers of ^{3}H-dihydro-MC-LR were investigated, whereas one was demonstrated to be taken up and distributed in various organs three to four times faster than the other epimer (Lindholm *et al.*, 1999). The latter observation was confirmed in conjunction with clearly distinguished variations regarding LD_{50} values of different MC congeners (LD_{50} of MC-LR: 50 µg/kg; LD_{50} of MC-RR: 500-800 µg/kg) (Zurawell *et al.*, 2005). These facts strongly support the assumption that already minimal structural variations lead to major alterations regarding uptake kinetics, organ distribution and subsequent toxicity (toxicokinetic). In addition, different animal strains, ages and exposure routs (p.o., i.v. and i.p.) used in those *in vivo* studies (Meriluoto *et al.*, 1990; Nishiwaki *et al.*, 1994; Hall *et al.*, 2008) together with varying time points of investigation (e.g. distribution of ^{3}H-dihydro-MC-LR 2.5 h vs. 48 h after exposure) resulted in limited comparability. It is very questionable weather or not single i.p or i.v. applications, except e.g. dialysis accidents, represent a potent risk scenario for humans. However, independent of the application route and differences regarding MC variants and epimers, these *in vivo* studies clearly indicate that MCs in general are capable for crossing the BBB/BCSFB and entering the brain.

The first study indicating brain abnormalities and therefore strongly suggesting a neurotoxic potential arising from a toxic *M. aeruginosa* extracts was performed in 1988 by Falconer *et al* (Falconer *et al.*, 1988). Female mice were orally exposed to one-fourth dilution of the toxic extract and mated with male mice from the same treatment. 10% of their 5-day old offspring clearly showed a reduced brain size with a 0.5 to 1.5 mm gap between the skull and the brain surface. One abnormal brain was investigated in more detail and indicated additionally an extensive damage in the outer region of the hippocampus, whereas no abnormalities were observed in the control group. Nevertheless, this study did not provide specific information about brain pathology in adult mice neither a closer characterization of the toxic extract used, especially in terms of different MC congeners and the employed concentration.

Next to the above described *in vivo* studies using mice as an model organisms, one recent human case report strongly suggest the neurotoxic potential of MCs. 131 patients of a hemodialysis unit in Caruaru, Brazil were accidentally exposed to MC-LR, -YR and –AR via dialysis water and 116 patients indicated immediate signs of acute neurotoxicity, including deafness, tinnitus and reversible blindness (Pouria *et al.*, 1998; Carmichael *et al.*, 2001; Azevedo *et al.*, 2002). Indeed, it is not clear whether the reported abnormal neurological symptoms in exposed patients derived indirectly from an effect of MC on the endothelium of the BBB with subsequent *in-situ* ischemia and inflammatory reactions, or directly via crossing the BBB targeting astrocytes, oligodendrocytes, microglia and/or especially neurons.

Preliminary in vitro evidence

To clarify the above described discrepancy of variable uptake kinetics and toxicity of different structural variants and to prove initial evidence for a direct brain-derived cell-type specific toxicity, primary murine whole brain cells (mWBCs, representing astrocytes, microglia and neurons) were employed and treated with single MC-LR, -LW and –LF. The result of this study thus suggests that MCs are capable for crossing mWBC membranes with subsequent covalent binding to ser/thr-specific PPs, significant PP inhibition and consequent toxicity already at 400 nM MC-LF (Chapter 2). Beside preliminary but fundamental *in vitro* evidence suggesting MCs as potent neurotoxins, the study failed to provide information about neuron-specific toxicity and pathological effects, downstream of PP inhibition (Chapter 2).

6.3. Involvement of organic anion transporting polypeptides in microcystin uptake

Cellular uptake of MCs has been demonstrated to occur exclusively via OATPs/Oatps (Chapter 1.3.2), while passive trans-membrane diffusion can be excluded (Eriksson *et al.*, 1990; Fischer *et al.*; 2005; Komatsu *et al.*, 2007). However, not all OATPs/Oatps are able to transport MCs (Fischer *et al.* 2005).

Consequently, the often stated hepatotoxicity and nephrotoxicity of MCs, is the result of a hepatic "first-pass-effect" and a subsequent renal "elimination effect" in organs having a high level of functionally expressed OATPs/Oatps capable for transporting MCs. So far, several OATPs/Oatps have been described in the human and rodent BBB/BCSFB (Table 1.3), including the MC-LR transporter OATP1A2 (Fischer *et al.*, 2005). Therefore, it was suggested that OATP1A2-mediated uptake of MCs across the human BBB is one of the prerequisites for the observed neurological symptoms of MC intoxicated patients after the fatal incident at the hemodialysis station in Caruaru (Dietrich and Hoeger, 2005; Fischer *et al.*, 2005).

It can be concluded that the expression profile of functional OATPs/Oatps in any given organ/tissue, capable for transporting MCs, appear to be absolute crucial, especially in conjunction with the observed hepato-, nephro- and neurotoxicity in MC intoxicated humans and animals.

6. Overall discussion

Presence of Oatps in primary neuronal cells

The conclusion mentioned above would denote on a cellular level that cell-type specific toxicity of MCs presuppose the presence of functional OATPs/Oatps in cell membranes, able for taken up MCs. Cellular uptake of MCs was demonstrated for OATP1B1, OATP1B3, mOatp1b2 located at the sinusoidal (basolateral) membrane of human/rat/mouse hepatocytes (Meier and Stieger, 2002; Fischer *et al.*, 2005; Letschert *et al.*, 2005) and OATP1A2 expressed e.g. at the luminal membrane of human endothelial cells of the BBB (Bronger *et al.*, 2005; Fischer *et al.*, 2005).

The latter observation could also be confirmed in preliminary experiments for primary mWBCs, expressing five of six tested mOatps on the mRNA level and additionally mOatp1b2, known to transport MC-LR, at the protein level (Chapter 2). However, mOatp function and their involvement in MCs uptake could only be demonstrated indirectly, simply to avoid the above described disadvantages of ^{3}H-dihydro-MC-LR (toxicokinetics). Therefore, the well characterized OATP/Oatp substrates taurocholate and bromosulfophthalein (van Montfoort *et al.*, 2003; Konig *et al.*, 2006; Meyer Zu Schwabedissen *et al.*, 2009) were employed and interestingly but not surprising, MC induced toxicity was significantly reduced in mWBC in the presence of the employed competitor.

Due to the scarcity of primary human neurons, primary murine cerebellar granule cells (mCGNs, representing more than 95% neurons), were used for mOatp investigation. PCR analysis, revealed the presence of all known mOatps at the mRNA level, except three members of the mOatp6 family (Figure 3.1). The expression of MC-LR transporter mOatp1b2, could be further confirmed at the protein level (Figure 3.2). Indeed, the presence of mOatp mRNA and even their protein expression do not devote that these transporters are functional. Uptake kinetics were performed using the well characterized OATP/Oatp substrates taurocholate and estrone sulfate (van Montfoort *et al.*, 2003; Konig *et al.*, 2006; Meyer Zu Schwabedissen *et al.*, 2009), thereby confirming the functionality, at least for mOatps, capable for transporting either one but most likely both substrates. This is due to the fact that all known MC-LR transporting OATPs/Oatps mediate the uptake of taurocholate and estrone sulfate (van Montfoort *et al.*, 2003; Konig *et al.*, 2006; Meyer Zu Schwabedissen *et al.*, 2009). Therefore it has to be assumed that these mOatps expressed in primary mCGNs are also capable of transporting MCs. Uptake was indirectly confirmed for all tested MC congeners, however only significant for MC-LF. More efficient transport of MC-LF was already reported previously in OATP1B1 and OATP1B3 transfected HeLa (Monks *et al.*, 2007) and HEK293 cells (Fischer *et al.*, in press).

It can be concluded that the neuronal uptake of MCs into primary murine neurons is governed by at least taurocholate and estrone sulfate transporting mOatps in general and specifically true, if not exclusively, by mOatp1b2. In addition, neuronal uptake is strongly dependent on the MC congener.

6. Overall discussion

Based on the presence of at least OATP3A1 in membranes of human glia cells and neurons (Huber et al., 2007), more than 60% amino acid identities between human and rodent OATP/Oatp subfamilies as well as up to 97% amino acid identities between the orthologues OATP3A1 and Oatp3a1 (Adachi et al., 2003; Hagenbuch and Meier, 2004) with similar to equal substrate preferences, it is therefore likely that MCs are capable for crossing membranes of human neurons as well. If the latter suggestion holds true, it would thus indicate similar MC uptake capabilities into human neurons compared with primary mCGNs. Thus, primary murine neurons would indicate an ideal cell system to study MC induced neurotoxicity on a molecular level.

6.4. Microcystin congener-dependent and neuron-specific toxicity

It was already mentioned above, that MC induced cell-type specific toxicity mainly depends on the expression profile and function of mOatps capable for transporting MCs. The latter was shown indirectly in preliminary experiments using primary mWBCs (Chapter 2). Interestingly, the observed toxicity in WBC was strongly dependent of the employed MC congener. MC-LF demonstrated already at 400 nM a significant reduction of cell viability, which was also true for MC-LW and MC-LR but only at ≥3 µM without providing information about neuron-specific toxicity and pathological effects, downstream of PP inhibition.

As we could elucidate that primary murine neurons express functional mOatps, including the MC-LR transporting mOatp1b2 (Chapter 3), it was assumed that mCGNs indicate similar MC congener dependent neurotoxic effects than observed in mWBCs. Already at non-cytotoxic concentrations of 500 nM MC-LF, 1 µM MC-LW and ~3 µM MC-LR, the neuronal cytoskeleton was disrupted, demonstrated by a significant reduction of the neurite length (Figure 4.6 and Figure 4.7). This process was most likely, if not exclusively due to pathological hyperphosphorylation of the microtubule associated Tau protein (Figure 4.5), a direct consequence of PP inhibition (Table 4.1). As already mentioned (Chapter 1.3.2.) it could be demonstrated in various cell-types that MC induced alterations in intermediate filaments and microfilaments as well (Wickstrom et al., 1995; Khan et al., 1996). Consequently it can not be excluded that other proteins e.g. cytokeratin8/18 (Ohta et al., 1992; Toivola and Eriksson, 1999) and subsequent cytoskeleton structures are negatively effected beside the observed microtubule breakdown after MC exposure (Figure 4.6).

At higher concentrations similar MC congener differences were demonstrated in primary mCGNs for caspase-3/7 mediated apoptosis (Chapter 4). Indeed, the latter finding was only true for MC-LF at ≥3 µM and for MC-LW at ≥5 µM, whereas MC-LR was not able to induce apoptosis at concentrations up to 10 µM (except a slight induction of caspase-3/7 at 10 µM). In addition, MC congener dependent necrosis in cultured mCGNs was tested in preliminary experiments and can be excluded. This is especially true in conjunction with the observed activation of caspases, representing a classic hallmark of "real" apoptosis i.e programmed cell death (Leist and Jaattela, 2001), and by more recent findings demonstrating MCs to induce apoptosis (Fladmark et al., 2002; Fu et al., 2005; Teneva et al., 2005; Komatsu et al., 2007). Interestingly, it could be elucidated that a 40% reduction of the neurite length was required for triggering the apoptotic pathway. Consequently, MC-LR did not reduce the neurite length below that threshold, therefore unable to undergo cell death in all the tested time and concentration experiments.

However, MC-LR, -LW, and –LF were recently demonstrated to have comparable PP-inhibitive capabilities using recombinant PPs (Monks et al., 2007; Fischer et al., in press) and MC-LR and MC-LF incubated brain homogenates (Table 4.1). Therefore it can be concluded that MC-LF and to a less extent MC-LW were transported more efficiently into primary murine neurons and thus reached intracellular PP-inhibition more quickly than did MC-LR. Consequently, MC-LF indicated the most potent neurotoxic MC congener among the tested variants already at low concentrations with similar characteristics than observed in AD. If these observations, including the higher toxicity of MC-LF, can be extrapolated to the human, they not only explain the observed neurotoxicity in the Caruaru accident, but also highlights the problems of using only one MC congener, MC-LR (WHO, 1999; Grosse et al., 2006), as the basis for microcystin related human risk assessment.

6.5. Assessment of human health risk, arising from microcystin congener differences and their neurotoxic potential

The guidance values proposed for MCs in drinking water by the WHO and for BGAS by the Oregon State Department of Health as well as general human exposure scenarios and consequent risk derived from MCs were more recently critically reviewed by Dietrich and Hoeger (2005) and Dietrich et al. (2008) and revealed overt scarcities. This is due to following reasons.

- The MC guidance values were derived from only one single MC congener, namely MC-LR, while at least 80 other MC congeners are for the most part unvalued.
- The MC guidance values were calculated from a TDI value based on a NOEL which was derived only from MC-LR induced hepatotoxicity whereas new knowledge provides largely unconsidered toxicity in other organs like the kidney and brain.
- The MC guidance values were calculated for an international adult unit (~60 kg), therefore providing insufficient protection for children.
- The inadequacies of assessing realistic MC exposure scenario in humans via contaminated drinking water, BGAS and/or recreational activities.

6. Overall discussion

- The inadequacies of assessing realistic MC uptake routes in humans (oral uptake and/or inhalation of MCs).
- The inadequacies of assessing realistic MC exposures scenarios in humans via chronic low dose exposure vs. acute high dose intoxication.

Based on the fact that MC-LF was demonstrated to be more neurotoxic *in vitro* (Chapter 2 and 4) and additionally more cytotoxic in OATP1B1 and OATP1B3 transfected HeLa (Monks *et al.*, 2007) and HEK293 cells (Fischer *et al.*, in press) it would likely indicate that MC-LF represents a more toxic variant *in vivo*. If the latter suggestion is also true for humans, this would indicate an underestimated MC risk assessment, which is presently based on MC-LR.

In addition, the *in vitro* data presented here in conjunction with the more recent *in vivo* studies (Table 1.4) clearly exhibit, brain pathology (Falconer *et al.*, 1988), memory loss (Maidana *et al.*, 2006) as well as neuron specific toxicity (Chapter 2 and 4). This could especially be true for humans in chronic low dose scenarios (e.g. via contaminated drinking water and/or BGAS) as low non-cytotoxic MC-LF and MC-LW concentrations revealed a significant neurite degeneration. As the NOEL (see above) is based on liver pathology of only MC-LR exposed mice without taking other MC congeners as well as pathology of other organs into closer considerations, it shoud be considered that MC guidance values are misguided and could be unreasonable.

6.6. Conclusions and future perspectives

It can be concluded that the observed MC congener differences are due to varying uptake capabilities mediated by mOatps. Therefore it seems to be absolute crucially to obtain a more in depth knowledge about MC congener dependent uptake by single Oatps/OATPs using e.g. stably transfected cell-lines or single Oatp/OATP expressing *Xenopus leavis* oocytes. Such studies would allow a better estimation of the toxicokinetic properties for individual MC congeners. Among the three tested MC congeners, only MC-LF and MC-LW are capable of inducing *in vitro* neurodegeneration at moderate to high concentrations with similar cellular characteristics as those observed in Alzheimer's Disease. As individual MC congeners differ strongly in their potential neurotoxicity, the current human risk assessment, based solely on MC-LR, may need a through revision. However, MC congener differences as well as the more subtle neurotoxicity of MC-LF and MC-LW at low non-cytotoxic concentrations need to be further investigated *in vivo*. If the latter holds true, MCs, and especially MC-LF would indicate a new and not yet carefully considered risk for humans and animals.

7. References

Abe, T., Kakyo, M., Tokui, T., Nakagomi, R., Nishio, T., Nakai, D., Nomura, H., Unno, M., Suzuki, M., Naitoh, T., Matsuno, S., and Yawo, H. (1999). Identification of a novel gene family encoding human liver-specific organic anion transporter LST-1. *J Biol Chem* **274**, 17159-17163.

Adachi, H., Suzuki, T., Abe, M., Asano, N., Mizutamari, H., Tanemoto, M., Nishio, T., Onogawa, T., Toyohara, T., Kasai, S., Satoh, F., Suzuki, M., Tokui, T., Unno, M., Shimosegawa, T., Matsuno, S., Ito, S., and Abe, T. (2003). Molecular characterization of human and rat organic anion transporter OATP-D. *Am J Physiol Renal Physiol* **285**, F1188-1197.

Adams, D. G., and Carr, N. G. (1981). The developmental biology of heterocyst and akinete formation in cyanobacteria. *Crit Rev Microbiol* **9**, 45-100.

Adams, D. G., and Duggan, P. S. (2008). Cyanobacteria-bryophyte symbioses. *J Exp Bot* **59**, 1047-1058.

Allen, M. M. (1984). Cyanobacterial cell inclusions. *Annu Rev Microbiol* **38**, 1-25.

Angeletti, R. H., Novikoff, P. M., Juvvadi, S. R., Fritschy, J. M., Meier, P. J., and Wolkoff, A. W. (1997). The choroid plexus epithelium is the site of the organic anion transport protein in the brain. *Proc Natl Acad Sci U S A* **94**, 283-286.

Annadotta, H., Cronberg, G., Lawton, L., Hansson, H.-B., Göthe, U., and Skulberg, O. M. (2001). An Extensive Outbreak of Gastroenteritis Associated with the Toxic Cyanobacterium *Planktothrix agardhii* (Oscillatoriales, Cyanophyceae) in Scania, South Sweden. In *Cyanotoxins* (I. Chorus, Ed.), pp. 200-208. Springer, Berlin, Heidelberg, New York.

Awramik, S. M. (1992). The oldest records of photosynthesis. *Photosynth Res* **33**, 75-89.

Azevedo, S. M. F. O., Evans, W. R., Carmichael, W. W., and Namikoshi, M. (1994). First report of microcystins from a Brazilian isolate of the cyanobacterium Microcystis aeruginosa. *Journal of Applied Phycology* **6**, 261-265.

Azevedo, S. M., Carmichael, W. W., Jochimsen, E. M., Rinehart, K. L., Lau, S., Shaw, G. R., and Eaglesham, G. K. (2002). Human intoxication by microcystins during renal dialysis treatment in Caruaru-Brazil. *Toxicology* **181-182**, 441-446.

Backer, L. C., Carmichael, W., Kirkpatrick, B., Williams, C., Irvin, M., Zhou, Y., Johnson, T. B., Nierenberg, K., Hill, V. R., Kieszak, S. M., and Cheng, Y. S. (2008). Recreational exposure to low concentrations of microcystins during an algal bloom in a small lake. *Mar Drugs* **6**, 389-406.

Bagu, J. R., Sykes, B. D., Craig, M. M., and Holmes, C. (1997). A molecular basis for different interactions of marine toxins with protein phosphatase-1 - Molecular models for bound motuporin, microcystins, okadaic acid, and calyculin A. *Journal of Biological Chemistry* **272**, 5087-5097.

Batista, T., de Sousa, G., Suput, J. S., Rahmani, R., and Suput, D. (2003). Microcystin-LR causes the collapse of actin filaments in primary human hepatocytes. *Aquat Toxicol* **65**, 85-91.

Beardall, J., Allen, D., Bragg, J., Finkel, Z. V., Flynn, K. J., Quigg, A., Rees, T. A., Richardson, A., and Raven, J. A. (2009). Allometry and stoichiometry of unicellular, colonial and multicellular phytoplankton. *New Phytol* **181**, 295-309.

Beasley, V. R., Lovell, R. A., Holmes, K. R., Walcott, H. E., Schaeffer, D. J., Hoffmann, W. E., and Carmichael, W. W. (2000). Microcystin-LR decreases hepatic and renal perfusion, and causes circulatory shock, severe hypoglycemia, and terminal hyperkalemia in intravascularly dosed swine. *Journal of Toxicology and Environmental Health A* **61**, 281-303.

Beattie, K. A., Kaya, K., Sano, T., and Codd, G. A. (1998). Three dehydrobutyrine-containing microcystins from Nostoc. *Phytochemistry (Oxford)* **47**, 1289-1292.

Bergwerk, A. J., Shi, X., Ford, A. C., Kanai, N., Jacquemin, E., Burk, R. D., Bai, S., Novikoff, P. M., Stieger, B., Meier, P. J., Schuster, V. L., and Wolkoff, A. W. (1996). Immunologic distribution of an organic anion transport protein in rat liver and kidney. *Am J Physiol* **271**, G231-238.

Berman-Frank, I., Lundgren, P., and Falkowski, P. (2003). Nitrogen fixation and photosynthetic oxygen evolution in cyanobacteria. *Res Microbiol* **154**, 157-164.

Berry, J. P., Gantar, M., Perez, M. H., Berry, G., and Noriega, F. G. (2008). Cyanobacterial toxins as allelochemicals with potential applications as algaecides, herbicides and insecticides. *Mar Drugs* **6**, 117-146.

Bezvenyuk, Z., Salminen, A., and Solovyan, V. (2000). Excision of DNA loop domains as a common step in caspase-dependent and -independent types of neuronal cell death. *Brain Res Mol Brain Res* **81**, 191-196.

Boichenko, V. A. (2004). Photosynthetic units of phototrophic organisms. *Biochemistry (Mosc)* **69**, 471-484.

Bologa, L., Joubert, R., Bisconte, J. C., Margules, S., Deugnier, M. A., Derbin, C., and Herschkowitz, N. (1983). Development of immunologically identified brain cells in culture: quantitative aspects. *Exp Brain Res* **53**, 163-167.

Botha, N., Gehringer, M. M., Downing, T. G., van de Venter, M., and Shephard, E. G. (2004). The role of microcystin-LR in the induction of apoptosis and oxidative stress in CaCo2 cells. *Toxicon* **43**, 85-92.

Botha, N., van de Venter, M., Downing, T. G., Shephard, E. G., and Gehringer, M. M. (2004a). The effect of intraperitoneally administered microcystin-LR on the gastrointestinal tract of Balb/c mice. *Toxicon* **43**, 251-254.

Bouaicha, N., and Maatouk, I. (2004). Microcystin-LR and nodularin induce intracellular glutathione alteration, reactive oxygen species production and lipid peroxidation in primary cultured rat hepatocytes. *Toxicology Letters* **148**, 53-63.

Boudreau, R. T., and Hoskin, D. W. (2005). The use of okadaic acid to elucidate the intracellular role(s) of protein phosphatase 2A: lessons from the mast cell model system. *Int Immunopharmacol* **5**, 1507-1518.

7. References

Boudreau, R. T., Conrad, D. M., and Hoskin, D. W. (2007). Apoptosis induced by protein phosphatase 2A (PP2A) inhibition in T leukemia cells is negatively regulated by PP2A-associated p38 mitogen-activated protein kinase. *Cell Signal* **19**, 139-151.

Bradford, M. M. (1976). A rapid and sensitive method for the quantification of mirogram quantities of protein utilizing the principle of protein-dye binding. *Analytical Biochemistry* **72**, 248-254.

Briand, J. F., Jacquet, S., Bernard, C., and Humbert, J. F. (2003). Health hazards for terrestrial vertebrates from toxic cyanobacteria in surface water ecosystems. *Vet Res* **34**, 361-377.

Brittain, S., Mohamed, Z. A., Wang, J., Lehmann, V. K., Carmichael, W. W., and Rinehart, K. L. (2000). Isolation and characterization of microcystins from a river Nile strain of *Oscillatoria tenuis* Agardh ex Gomont. *Toxicon* **38**, 1759-1771.

Bronger, H., Konig, J., Kopplow, K., Steiner, H. H., Ahmadi, R., Herold-Mende, C., Keppler, D., and Nies, A. T. (2005). ABCC drug efflux pumps and organic anion uptake transporters in human gliomas and the blood-tumor barrier. *Cancer Res* **65**, 11419-11428.

Burns, J. W. (2004). Cyanotoxin in Floridas (USA) surface waters: considerations for water supply planning. In ICTV 6th, Bergen, Norway, pp4.

Callaghan, T. V., Bjorn, L. O., Chernov, Y., Chapin, T., Christensen, T. R., Huntley, B., Ims, R. A., Johansson, M., Jolly, D., Jonasson, S., Matveyeva, N., Panikov, N., Oechel, W., Shaver, G., Elster, J., Henttonen, H., Laine, K., Taulavuori, K., Taulavuori, E., and Zockler, C. (2004). Biodiversity, distributions and adaptations of Arctic species in the context of environmental change. *Ambio* **33**, 404-417.

Carbis, C. R., Simons, J. A., Mitchell, G. F., Anderson, J. W., and McCauley, I. (1994). A biochemical profile for predicting the chronic exposure of sheep to Microcystis aeruginosa, an hepatotoxic species of blue-green alga. *Res Vet Sci* **57**, 310-316.

Carbis, C. R., Waldron, D. L., Mitchell, G. F., Anderson, J. W., and McCauley, I. (1995). Recovery of hepatic function and latent mortalities in sheep exposed to the blue-green alga Microcystis aeruginosa. *Vet. Record* **137**, 12-15.

Carmichael, W. W., and Biggs, D. F. (1978). Muscle sensitivity differences in two avian species to anatoxin-a produced by the freshwater cyanophyte Anabaena flos-aquae NRC-44-1. *Canadian Journal of Zoology* **56**, 510-512.

Carmichael, W. W., Biggs, D. F., and Peterson, M. A. (1979). Pharmacology of anatoxin-a, produced by the freshwater cyanophyte Anabaena flos-aquae NRC-44-1. *Toxicon* **17**, 229-236.

Carmichael, W. W., Beasley, V., Bunner, D. L., Eloff, J. N., Falconer, I., Gorham, P., Harada, K., Krishnamurthy, T., Yu, M. J., Moore, R. E., and et al. (1988). Naming of cyclic heptapeptide toxins of cyanobacteria (blue-green algae). *Toxicon* **26**, 971-973.

Carmichael, W. W. (1997). The Cyanotoxins. *Advances in Botanical Research* **27**, 211-256.

Carmichael, W. W., Azevedo, S. M., An, J. S., Molica, R. J., Jochimsen, E. M., Lau, S., Rinehart, K. L., Shaw, G. R., and Eaglesham, G. K. (2001). Human fatalities from cyanobacteria: chemical and biological evidence for cyanotoxins. *Environmental Health Perspectives* **109**, 663-668.

Carmichael, W. W. (2001a). Health Effects of Toxin-Producing Cyanobacteria: "The CyanoHABs". *Human and Ecological Risk Assessment* **7**, 1393-1407.

Carmichael, W. (2008). A world overview--one-hundred-twenty-seven years of research on toxic cyanobacteria--where do we go from here? *Adv Exp Med Biol* **619**, 105-125.

Carpenter, E. J., Bergman, B., Dawson, R., Siddiqui, P. J., Soderback, E., and Capone, D. G. (1992). Glutamine synthetase and nitrogen cycling in colonies of the marine diazotrophic cyanobacteria Trichodesmium spp. *Appl Environ Microbiol* **58**, 3122-3129.

Castenholz, R. W., and Waterbury, J. B. (1989). Oxygenic photosynthetic bacteria. Group 1. Cyanobacteria. In *Bergey's Manual of Systematic Bacteriology* (J. T. Stanley, M. P. Bryant, N. Pfennig, and J. G. Holt, Eds.), pp. 1710-1806, Baltimore.

Cazenave, J., Wunderlin, D. A., de Los Angeles Bistoni, M., Ame, M. V., Krause, E., Pflugmacher, S., and Wiegand, C. (2005). Uptake, tissue distribution and accumulation of microcystin-RR in Corydoras paleatus, Jenynsia multidentata and Odontesthes bonariensis. A field and laboratory study. *Aquat Toxicol* **75**, 178-190.

Cecchelli, R., Berezowski, V., Lundquist, S., Culot, M., Renftel, M., Dehouck, M. P., and Fenart, L. (2007). Modelling of the blood-brain barrier in drug discovery and development. *Nat Rev Drug Discov* **6**, 650-661.

Chen, J., Xie, P., Guo, L., Zheng, L., and Ni, L. (2005). Tissue distributions and seasonal dynamics of the hepatotoxic microcystins-LR and -RR in a freshwater snail (Bellamya aeruginosa) from a large shallow, eutrophic lake of the subtropical China. *Environ Pollut* **134**, 423-430.

Cheng, X., Maher, J., Chen, C., and Klaassen, C. D. (2005). Tissue distribution and ontogeny of mouse organic anion transporting polypeptides (oatps). *Drug Metab Dispos* **33**, 1062-1073.

Chorus, I., and Bartram, J. (1999). *Toxic Cyanobacteria in Water: A Guide to Their Puplic Health Consequences, Monitoring and Management*. Routeledge, London UK.

Claeyssens, S., Chedeville, A., and Lavoinne, A. (1993). Inhibition of protein phosphatases activates glucose-6-phosphatase in isolated rat hepatocytes. *FEBS Lett* **315**, 7-10.

Claeyssens, S., Francois, A., Chedeville, A., and Lavoinne, A. (1995). Microcystin-LR induced an inhibition of protein synthesis in isolated rat hepatocytes. *Biochemistry Journal* **306**, 693-696.

Clorinda Arias, N. S. P. D. B. S.-Z. (1993). Okadaic Acid Induces Early Changes in Microtubule-Associated Protein 2 and Phosphorylation Prior to Neurodegeneration in Cultured Cortical Neurons. *Journal of Neurochemistry* **61**, 673-682.

7. References

Codd, G. A., Steffensen, D. A., Burch, M. D., and Baker, P. D. (1994). Toxic blooms of cyanobacteria in Lake Alexandrina, South Australia - learning from history. *Australian Journal of Marine and Freshwater Research* **45**, 731-736.

Codd, G. (1995). Cyanobacterial toxins: occurrence, properties and biological significance. *Water Science & Technology* **32**, 149-156.

Cooney, R. P., Pantos, O., Le Tissier, M. D., Barer, M. R., O'Donnell, A. G., and Bythell, J. C. (2002). Characterization of the bacterial consortium associated with black band disease in coral using molecular microbiological techniques. *Environ Microbiol* **4**, 401-413.

Cox, P. A., Banack, S. A., and Murch, S. J. (2003). Biomagnification of cyanobacterial neurotoxins and neurodegenerative disease among the Chamorro people of Guam. *Proc Natl Acad Sci U S A* **100**, 13380-13383.

Cox, P. A., Banack, S. A., Murch, S. J., Rasmussen, U., Tien, G., Bidigare, R. R., Metcalf, J. S., Morrison, L. F., Codd, G. A., and Bergman, B. (2005). Diverse taxa of cyanobacteria produce beta-N-methylamino-L-alanine, a neurotoxic amino acid. *Proc Natl Acad Sci U S A* **102**, 5074-5078.

Craig, M., Luu, H. A., McCready, T. L., Williams, D., Andersen, R. J., and Holmes, C. (1996). Molecular mechanisms underlying the interaction of motuporin and microcystins with type-1 and type-2A protein phosphatases. *Biochemistry and Cell Biology - Biochimie et Biologie Cellulaire* **74**, 569-578.

Cribbs, D. H., Poon, W. W., Rissman, R. A., and Blurton-Jones, M. (2004). Caspase-Mediated Degeneration in Alzheimer's Disease. *Am J Pathol* **165**, 353-355.

Cruz-Aguado, R., Winkler, D., and Shaw, C. A. (2006). Lack of behavioral and neuropathological effects of dietary beta-methylamino-L-alanine (BMAA) in mice. *Pharmacol Biochem Behav* **84**, 294-299.

Dahlman, L., Persson, J., Palmqvist, K., and Nasholm, T. (2004). Organic and inorganic nitrogen uptake in lichens. *Planta* **219**, 459-467.

de Figueiredo, D. R., Azeiteiro, U. M., Esteves, S. M., Goncalves, F. J., and Pereira, M. J. (2004). Microcystin-producing blooms--a serious global public health issue. *Ecotoxicol Environ Saf* **59**, 151-163.

DeMott, W. R., Zhang, Q.-X., and Carmichael, W. W. (1991). Effects of toxic cyanobacteria and purified toxins on the survival and feeding of a copepod and three species of Daphnia. *Limnology and Oceanography* **36**, 1346-1357.

DeVries, S. E., Galey, F. D., Namikoshi, M., and Woo, J. C. (1993). Clinical and pathologic findings of blue-green algae (*Microcystis aeruginosa*) intoxication in a dog. *Journal of Veterinary Diagnostic Investigation* **5**, 403-408.

DeVries, S. E., Namikoshi, M., Galey, F. D., Merritt, J. E., Rinehart, K. L., and Beasley, V. R. (1993). Chemical study of the hepatotoxins from Microcystis aeruginosa collected in California. *J Vet Diagn Invest* **5**, 409-412.

Dietrich, D., and Hoeger, S. (2005). Guidance values for microcystins in water and cyanobacterial supplement products (blue-green algal supplements): a reasonable or misguided approach? *Toxicol Appl Pharmacol* **203**, 273-289.

Dietrich, D. R., Fischer, A., Michel, C., and Hoeger, S. J. (2008). Toxin mixture in cyanobacterial blooms--a critical comparison of reality with current procedures employed in human health risk assessment. *Adv Exp Med Biol* **619**, 885-912.

Dillenberg, H. O., and Dehnel, M. K. (1960). Toxic waterbloom in Saskatchewan, 1959. *Canadian Medical Association Journal* **83**, 1151-1154.

Ding, W. X., Shen, H. M., and Ong, C. N. (2002). Calpain activation after mitochondrial permeability transition in microcystin-induced cell death in rat hepatocytes. *Biochem Biophys Res Commun* **291**, 321-331.

Ding, W. X., and Nam Ong, C. (2003). Role of oxidative stress and mitochondrial changes in cyanobacteria-induced apoptosis and hepatotoxicity. *FEMS Microbiol Lett* **220**, 1-7.

Dittmann, E., and Wiegand, C. (2006). Cyanobacterial toxins - occurrence, biosynthesis and impact on human affairs. *Molecular Nutrition & Food Research* **50**, 7-17.

Duncan, M. W., and Marini, A. M. (2006). Debating the cause of a neurological disorder. *Science* **313**, 1737.

Edwards, C., Beattie, K. A., Scrimgeour, C. M., and Codd, G. A. (1992). Identification of anatoxin-A in benthic cyanobacteria (blue-green algae) and in associated dog poisonings at Loch Insh, Scotland. *Toxicon* **30**, 1165-1175.

Edwards, C., Graham, D., Fowler, N., and Lawton, L. A. (2008). Biodegradation of microcystins and nodularin in freshwaters. *Chemosphere* **73**, 1315-1321.

Eriksson, J. E., Hägerstrand, H., and Isomaa, B. (1987). Cell selective cytotoxicity of a peptide toxin from the cyanobacterium Microcystis aeruginosa. *Biochimica et Biophysica Acta* **930**, 304-310.

Eriksson, J. E., Paatero, G. I. L., Meriluoto, J. A. O., Codd, G. A., Kass, G. E. N., Nicotera, P., and Orrenius, S. (1989). Rapid microfilament reorganization induced in isolated rat hepatocytes by microcystin-LR, a cyclic peptide toxin. *Experimental Cell Research* **185**, 86-100.

Eriksson, J. E., Gronberg, L., Nygard, S., Slotte, J. P., and Meriluoto, J. A. (1990). Hepatocellular uptake of 3H-dihydromicrocystin-LR, a cyclic peptide toxin. *Biochim Biophys Acta* **1025**, 60-66.

Eriksson, J. E., Toivola, D., Meriluoto, J. A., Karaki, H., Han, Y. G., and Hartshorne, D. (1990a). Hepatocyte deformation induced by cyanobacterial toxins reflects inhibition of protein phosphatases. *Biochem Biophys Res Commun* **173**, 1347-1353.

Ernst, B., Hitzfeld, B., and Dietrich, D. R. (2000). Detection of cyanobacterial toxins in whitefish (*Coregonus lavaretus*) from Lake Ammersee. *Toxicological Sciences* **54**, 330.

Ernst, B., Hitzfeld, B., and Dietrich, D. (2001). Presence of *Planktothrix* sp. and cyanobacterial toxins in Lake Ammersee, Germany and their impact on whitefish (*Coregonus lavaretus* L.). *Environmental Toxicology* **16**, 483-488.

Ernst, B. (2001a). Cyanobakterien auf dem Vormarsch. In *GEO*, pp. 217-219.

7. References

Falconer, I. R., Beresford, A. M., and Runnegar, M. T. (1983). Evidence of liver damage by toxin from a bloom of the blue-green alga, Microcystis aeruginosa. *Med J Aust* **1**, 511-514.

Falconer, I. R., Smith, J. V., Jackson, A. R., Jones, A., and Runnegar, M. T. (1988). Oral toxicity of a bloom of the Cyanobacterium *Microcystis aeruginosa* administered to mice over periods up to 1 year. *Journal of Toxicology and Environmental Health* **24**, 291-305.

Falconer, I. R., and Yeung, D. S. K. (1992). Cytoskeletal changes in hepatocytes induced by *Microcystis* toxins and their relation to hyperphosphorylation of cell proteins. *Chemico-Biological Interactions* **81**, 181-196.

Falconer, I. R. (1999). An overview of problems caused by toxic blue-green algae (cyanobacteria) in drinking and recreational water. *Environmental Toxicology* **14**, 5-12.

Falconer, I. R., and Humpage, A. R. (2005). Health risk assessment of cyanobacterial (blue-green algal) toxins in drinking water. *Int J Environ Res Public Health* **2**, 43-50.

Falconer, I. R. (2005a). Is there a human health hazard from microcystins in the drinking water supply? *Acta Hydrochimica Et Hydrobiologica* **33**, 64-71.

Falconer, I. R. (2008). Health effects associated with controlled exposures to cyanobacterial toxins. *Adv Exp Med Biol* **619**, 607-612.

Fastner, J., Flieger, I., and Neumann, U. (1998). Optimised extraction of microcystins from field samples - a comparison of different solvents and procedures. *Water Research* **32**, 3177-3181.

Fawell, J., James, C., and James, H. (1994). Toxins from blue-green algae: Toxicological assessment of Microcystin-LR and a method for its determination in water. Foundation for Water Research, Marlow.

Fawell, J. K., Mitchell, R. E., Everett, D. J., and Hill, R. E. (1999). The toxicity of cyanobacterial toxins in the mouse: I microcystin-LR. *Hum Exp Toxicol* **18**, 162-167.

Fawell, J. K., Mitchell, R. E., Hill, R. E., and Everett, D. J. (1999a). The toxicity of cyanobacterial toxins in the mouse: II anatoxin-a. *Hum Exp Toxicol* **18**, 168-173.

Fay, P. (1992). Oxygen relations of nitrogen fixation in cyanobacteria. *Microbiol Rev* **56**, 340-373.

Feurstein, D., Holst, K., Fischer, A., and Dietrich, D. R. (2009). Oatp-associated uptake and toxicity of microcystins in primary murine whole brain cells. *Toxicol Appl Pharmacol* **234**, 247-255.

Fialkowska, E., and Pajdak-Stos, A. (1997). Inducible defence against ciliate grazer, *Pseudomicrothorax dubius*, in two strains of *Phormidium* (cyanobacteria). *Proc. R. Soc. London B* **264**, 937-941.

Fialkowska, E., and Pajdak-Stos, A. (2002). Dependence of cyanobacteria defense mode on grazer pressure. *Aquatic Microbial Ecology* **27**, 149-157.

Fischer, W. J., and Dietrich, D. R. (2000). Pathological and biochemical characterization of microcystin-induced hepatopancreas and kidney damage in carp (Cyprinus carpio). *Toxicol Appl Pharmacol* **164**, 73-81.

Fischer, W. J., and Dietrich, D. R. (2000a). Toxicity of the cyanobacterial cyclic heptapeptide toxins microcystin- LR and -RR in early life-stages of the African clawed frog (*Xenopus laevis*). *Aquatic Toxicology* **49**, 189-198.

Fischer, W. J., Garthwaite, I., Miles, C. O., Ross, K. M., Aggen, J. B., Chamberlin, A. R., Towers, N. R., and Dietrich, D. R. (2001). Congener-independent immunoassay for microcystins and nodularins. *Environmental Science & Technology* **35**, 4849-4856.

Fischer, W. J., Altheimer, S., Cattori, V., Meier, P. J., Dietrich, D. R., and Hagenbuch, B. (2005). Organic anion transporting polypeptides expressed in liver and brain mediate uptake of microcystin. *Toxicol Appl Pharmacol* **203**, 257-263.

Fischer, A., Hoeger, S. J., Stemmer, K., Feurstein, D., Knobeloch D., Nüssler A. and Dietrich D.R. The role of organic anion transporting polypeptides (OATPs/SLCOs) for the toxicity of different microcystin congeners in vitro: a comparison of primary human hepatocytes and OATP-transfected HEK293 cells. (Toxicology and Applied Pharmacology, in press).

Fitzgeorge, R. B., Clark, S. A., and Keevil, C. W. (1994). Routes of intoxication. In *Detection methods for cyanobacterial toxins* (G. A. Codd, T. M. Jefferies, C. W. Keevil, and E. Potter, Eds.). Royal Society of Chemistry, Cambridge, UK.

Fladmark, K. E., Brustugun, O. T., Hovland, R., Boe, R., Gjertsen, B. T., Zhivotovsky, B., and Doskeland, S. O. (1999). Ultrarapid caspase-3 dependent apoptosis induction by serine/threonine phosphatase inhibitors. *Cell Death Differentiation* **6**, 1099-1108.

Fladmark, K. E., Brustugun, O. T., Mellgren, G., Krakstad, C., Boe, R., Vintermyr, O. K., Schulman, H., and Doskeland, S. O. (2002). Ca2+/calmodulin-dependent protein kinase II is required for microcystin-induced apoptosis. *J Biol Chem* **277**, 2804-2811.

Fleming, L. E., Rivero, C., Burns, J. L., Williams, C., Bean, J. A., Shea, K. A., and Stinn, J. (2002). Blue green algal (cyanobacterial) toxins, surface drinking water, and liver cancer in Florida. *Harmful Algae* **1**, 157-168.

Flores, E., Frias, J. E., Rubio, L. M., and Herrero, A. (2005). Photosynthetic nitrate assimilation in cyanobacteria. *Photosynth Res* **83**, 117-133.

Francis, G. (1878). Poisonous Australian lake. *Nature* **18**, 11-12.

Fu, W. Y., Chen, J. P., Wang, X. M., and Xu, L. H. (2005). Altered expression of p53, Bcl-2 and Bax induced by microcystin-LR in vivo and in vitro. *Toxicon* **46**, 171-177.

Fujiki, H. (1992). Is the inhibition of protein phosphatase 1 and 2A activities a general mechanism of tumor promotion in human cancer development? *Molecular Carcinogenesis* **5**, 91-94.

7. References

Fujiki, H., and Suganuma, M. (1993). Tumor promotion by inhibitors of protein phosphatases 1 and 2A: the okadaic acid class of compounds. *Advances in Cancer Research* **61**, 143-194.

Fujiki, H., Sueoka, E., and Suganuma, M. (1996). Carcinogenesis of Microcystins. In *Toxic Microcystis* (M. F. Watanabe, K. Harada, W. W. Carmichael, and H. Fujiki, Eds.), pp. 203-232. 1996 CRC Press Inc., Boca Raton, New York, London, Tokyo.

Gao, B., Stieger, B., Noe, B., Fritschy, J. M., and Meier, P. J. (1999). Localization of the organic anion transporting polypeptide 2 (Oatp2) in capillary endothelium and choroid plexus epithelium of rat brain. *J Histochem Cytochem* **47**, 1255-1264.

Gao, B., Hagenbuch, B., Kullak-Ublick, G. A., Benke, D., Aguzzi, A., and Meier, P. J. (2000). Organic anion-transporting polypeptides mediate transport of opioid peptides across blood-brain barrier. *J Pharmacol Exp Ther* **294**, 73-79.

Gao, B., Huber, R. D., Wenzel, A., Vavricka, S. R., Ismair, M. G., Reme, C., and Meier, P. J. (2005). Localization of organic anion transporting polypeptides in the rat and human ciliary body epithelium. *Exp Eye Res* **80**, 61-72.

Garcia-Pichel, F., Mechling, M., and Castenholz, R. W. (1994). Diel Migrations of Microorganisms within a Benthic, Hypersaline Mat Community. *Appl Environ Microbiol* **60**, 1500-1511.

Garcia-Pichel, F., Prufert-Bebout, L., and Muyzer, G. (1996). Phenotypic and phylogenetic analyses show Microcoleus chthonoplastes to be a cosmopolitan cyanobacterium. *Appl Environ Microbiol* **62**, 3284-3291.

Garcia-Pichel, F. (1998). Solar ultraviolet and the evolutionary history of cyanobacteria. *Orig Life Evol Biosph* **28**, 321-347.

Gehringer, M. M. (2004). Microcystin-LR and okadaic acid-induced cellular effects: a dualistic response. *FEBS Lett* **557**, 1-8.

Ghosh, S., Khan, S. A., Wickstrom, M., and Beasley, V. (1995). Effects of microcystin-LR on actin and the actin-associated proteins alpha-actinin and talin in hepatocytes. *Natural Toxins* **3**, 405-414.

Gilroy, D. J., Kauffman, K. W., Hall, R. A., Huang, X., and Chu, F. S. (2000). Assessing potential health risks from microcystin toxins in blue-green algae dietary supplements. *Environmental Health Perspectives* **108**, 435-439.

Goldberg, J., Huang, H. B., Kwon, Y. G., Greengard, P., Nairn, A. C., and Kuriyan, J. (1995). Three-dimensional structure of the catalytic subunit of protein serine/threonine phosphatase-1. *Nature* **376**, 745-753.

Gong, C. X., Singh, T. J., Grundke-Iqbal, I., and Iqbal, K. (1993). Phosphoprotein phosphatase activities in Alzheimer disease brain. *J Neurochem* **61**, 921-927.

Gong, C. X., Shaikh, S., Wang, J. Z., Zaidi, T., Grundke-Iqbal, I., and Iqbal, K. (1995). Phosphatase activity toward abnormally phosphorylated tau: decrease in Alzheimer disease brain. *J Neurochem* **65**, 732-738.

Gong, C. X., Liu, F., Grundke-Iqbal, I., and Iqbal, K. (2005). Post-translational modifications of tau protein in Alzheimer's disease. *Journal of Neural Transmission* **112**, 813-838.

Gorbushina, A. A. (2007). Life on the rocks. *Environ Microbiol* **9**, 1613-1631.

Gray, J. E., and Herson, D. S. (1976). Functional 70S hybrid ribosomes from blue-green algae and bacteria. *Arch Microbiol* **109**, 95-99.

Grosse, Y., Baan, R., Straif, K., Secretan, B., El Ghissassi, F., and Cogliano, V. (2006). Carcinogenicity of nitrate, nitrite, and cyanobacterial peptide toxins. *Lancet Oncol* **7**, 628-629.

Gulledge, B. M., Aggen, J. B., Eng, H., Sweimeh, K., and Chamberlin, A. R. (2003). Microcystin analogues comprised only of Adda and a single additional amino acid retain moderate activity as PP1/PP2A inhibitors. *Bioorg Med Chem Lett* **13**, 2907-2911.

Guo, H., Albrecht, S., Bourdeau, M., Petzke, T., Bergeron, C., and LeBlanc, A. C. (2004). Active caspase-6 and caspase-6-cleaved tau in neuropil threads, neuritic plaques, and neurofibrillary tangles of Alzheimer's disease. *Am J Pathol* **165**, 523-531.

Guven, B., and Howard, A. (2006). Modelling the growth and movement of cyanobacteria in river systems. *Sci Total Environ* **368**, 898-908.

Guzman, R. E., Solter, P. F., and Runnegar, M. T. (2003). Inhibition of nuclear protein phosphatase activity in mouse hepatocytes by the cyanobacterial toxin microcystin-LR. *Toxicon* **41**, 773-781.

Hagenbuch, B., and Meier, P. J. (2003). The superfamily of organic anion transporting polypeptides. *Biochim Biophys Acta* **1609**, 1-18.

Hagenbuch, B., and Meier, P. J. (2004). Organic anion transporting polypeptides of the OATP/ SLC21 family: phylogenetic classification as OATP/ SLCO superfamily, new nomenclature and molecular/functional properties. *Pflugers Arch* **447**, 653-665.

Hagenbuch, B. (2007). Cellular entry of thyroid hormones by organic anion transporting polypeptides. *Best Pract Res Clin Endocrinol Metab* **21**, 209-221.

Hall, A., Tibbetts, B. M., Rein, K. S. (2008). Microcystin-LR crosses the placenta in CD-1 mice. In: SOT 2008, Seattle, USA.

Hanger, D. P., Betts, J. C., Loviny, T. L., Blackstock, W. P., and Anderton, B. H. (1998). New phosphorylation sites identified in hyperphosphorylated tau (paired helical filament-tau) from Alzheimer's disease brain using nanoelectrospray mass spectrometry. *J Neurochem* **71**, 2465-2476.

Harada, K., Oshikata, M., Uchida, H., Suzuki, M., Kondo, F., Sato, K., Ueno, Y., Yu, S. Z., Chen, G., and Chen, G. C. (1996). Detection and identification of microcystins in the drinking water of Haimen City, China. *Natural Toxins* **4**, 277-283.

7. References

Harada, K.-i., and Ueno, Y. (1996a). Studies on microcystin contents in different drinking water in highly endemic area of liver cancer. *Chinese Journal of Preventive Medicine* **30**, 6-9.

Harding, W. R., Rowe, N., Wessels, J. C., Beattie, K. A., and Codd, G. A. (1995). Death of a dog attributed to the cyanobacterial (blue-green algal) hepatotoxin nodularin in South Africa. *Journal of the South African Veterinary Association* **66**, 256-259.

Hastie, C. J., Borthwick, E. B., Morrison, L. F., Codd, G. A., and Cohen, P. T. (2005). Inhibition of several protein phosphatases by a non-covalently interacting microcystin and a novel cyanobacterial peptide, nostocyclin. *Biochim Biophys Acta* **1726**, 187-193.

Heinze, R. (1999). Toxicity of the cyanobacterial toxin microcystin-LR to rats after 28 days intake with the drinking water. *Environmental Toxicology* **14**, 57-60.

Henriksen, P. (1996). Microcystin profiles and contents in Danish populations of cyanobacteria/blue-green algae as determined by HPLC. *Phycologica* **35**, 102-110.

Henriksen, P., Carmichael, W. W., An, J. S., and Moestrup, O. (1997). Detection of an anatoxin-a(s)-like anticholinesterase in natural blooms and cultures of Cyanobacteria/blue-green algae from Danish lakes and in the stomach contents of poisoned birds. *Toxicon* **35**, 901-913.

Hitzfeld, B. C., Lampert, C. S., Spaeth, N., Mountfort, D., Kaspar, H., and Dietrich, D. R. (2000). Toxin production in cyanobacterial mats from ponds on the McMurdo ice shelf, Antarctica. *Toxicon* **38**, 1731-1748.

Hitzfeld, B. C., Hoger, S. J., and Dietrich, D. R. (2000a). Cyanobacterial toxins: removal during drinking water treatment, and human risk assessment. *Environ Health Perspect* **108** Suppl 1, 113-122.

Hoeger, S. J. (2003). PhD thesis (http://nbn-resolving.de/urn:nbn:de:bsz:352-opus-10714).

Hoeger, S. J., Shaw, G., Hitzfeld, B. C., and Dietrich, D. R. (2004). Occurrence and elimination of cyanobacterial toxins in two Australian drinking water treatment plants. *Toxicon* **43**, 639-649.

Hoiczyk, E., and Hansel, A. (2000). Cyanobacterial cell walls: news from an unusual prokaryotic envelope. *J Bacteriol* **182**, 1191-1199.

Holtzman, D., Olson, J. E., DeVries, C., and Bensch, K. (1987). Lead toxicity in primary cultured cerebral astrocytes and cerebellar granular neurons. *Toxicol Appl Pharmacol* **89**, 211-225.

Honkanen, R. E., Zwiller, J., Moore, R. E., Daily, S. L., Khatra, B. S., Dukelow, M., and Boynton, A. L. (1990). Characterization of microcystin-LR, a potent inhibitor of type 1 and type 2A protein phosphatases. *Journal of Biological Chemistry* **265**, 19401-19404.

Hooser, S. B., Beasley, V. R., Lovell, R. A., Carmichael, W. W., and Haschek, W. M. (1989). Toxicity of microcystin LR, a cyclic heptapeptide hepatotoxin from Microcystis aeruginosa, to rats and mice. *Vet Pathol* **26**, 246-252.

Hooser, S. B., Waite, L. L., Beasley, V. R., Carmichael, W. W., Kuhlenschmidt, M., and Haschek, W. M. (1989). Microcystin-A induces morphologic and cytoskeletal hepatocyte changes in vitro. *Toxicon* **27**, 50-51.

Hooser, S., Beasley, V., Waite, L., Kuhlenschmidt, M., Carmicheal, W., and Haschek, W. (1991). Actin filament alterations in rat hepatocytes induced in vivo and in vitro by microcystin-LR, a hepatotoxin from the blue-green alga, *Microcystis aeruginosa*. *Veterinary Pathology* **28**.

Hooser, S. B. (2000). Fulminant hepatocyte apoptosis in vivo following microcystin-LR administration to rats. *Toxicologic Pathology* **28**, 726-733.

Hoppu, K. (2002). [Treatment of paracetamol intoxication. Correctly given antidote treatment will prevent liver damage]. *Duodecim* **118**, 187-191.

Huber, R. D., Gao, B., Sidler Pfandler, M. A., Zhang-Fu, W., Leuthold, S., Hagenbuch, B., Folkers, G., Meier, P. J., and Stieger, B. (2007). Characterization of two splice variants of human organic anion transporting polypeptide 3A1 isolated from human brain. *Am J Physiol Cell Physiol* **292**, C795-806.

Humpage, A. R., and Falconer, I. R. (1999). Microcystin-LR and liver tumor promotion: Effects on cytokinesis, ploidy, and apoptosis in cultured hepatocytes. *Environmental Toxicology* **14**, 61-75.

Humpage, A. R., Hardy, S. J., Moore, E. J., Froscio, S. M., and Falconer, I. R. (2000). Microcystins (cyanobacterial toxins) in drinking water enhance the growth of aberrant crypt foci in the mouse colon. *Journal of Toxicology and Environmental Health* **61**, 155-165.

Humpage, A. (2008). Toxin types, toxicokinetics and toxicodynamics. *Adv Exp Med Biol* **619**, 383-415.

Ibelings, B. W., and Chorus, I. (2007). Accumulation of cyanobacterial toxins in freshwater "seafood" and its consequences for public health: a review. *Environ Pollut* **150**, 177-192.

Iqbal, K., and G.-I, I. (2008). Alzheimer neurofibrillary degeneration: significance, etiopathogenesis, therapeutics and prevention. *Journal of Cellular and Molecular Medicine* **12**, 38-55.

Ito, E., Kondo, F., Terao, K., and Harada, K.-I. (1997). Neoplastic nodular formation in mouse liver induced by repeated intraperitoneal injections of microcystin-LR. *Toxicon* **35**, 1453-1457.

Ito, E., Kondo, F., and Harada, K. I. (1997a). Hepatic necrosis in aged mice by oral administration of microcystin-LR. *Toxicon* **35**, 231-239.

Ito, E., Kondo, F., and Harada, K. (2000). First report on the distribution of orally administered microcystin-LR in mouse tissue using an immunostaining method. *Toxicon* **38**, 37-48.

Jacquemin, E., Hagenbuch, B., Stieger, B., Wolkoff, A. W., and Meier, P. J. (1994). Expression cloning of a rat liver Na(+)-independent organic anion transporter. *Proc Natl Acad Sci U S A* **91**, 133-137.

7. References

Jochimsen, E. M., Carmichael, W. W., An, J. S., Cardo, D. M., Cookson, S. T., Holmes, C. E., Antunes, M. B., de Melo Filho, D. A., Lyra, T. M., Barreto, V. S., Azevedo, S. M., and Jarvis, W. R. (1998). Liver failure and death after exposure to microcystins at a hemodialysis center in Brazil. *N Engl J Med* **338**, 873-878.

Jones, G. J., and Orr, P. T. (1994). Release and degradation of microcystin following algicide treatment of a *Microcystis aeruginosa* bloom in a recreational lake, as determined by HPLC and protein phosphatase inhibition assay. *Water Research* **28**, 871-876.

Jones, G. J., Bourne, D. G., Blakeley, R. L., and Doelle, H. (1994a). Degradation of the cyanobacterial hepatotoxin microcystin by aquatic bacteria. *Natural Toxins* **2**, 228-235.

Jungblut, A. D., Hawes, I., Mountfort, D., Hitzfeld, B., Dietrich, D. R., Burns, B. P., and Neilan, B. A. (2005). Diversity within cyanobacterial mat communities in variable salinity meltwater ponds of McMurdo Ice Shelf, Antarctica. *Environmental Microbiology* **7**, 519-529.

Kanai, N., Lu, R., Bao, Y., Wolkoff, A. W., and Schuster, V. L. (1996). Transient expression of oatp organic anion transporter in mammalian cells: identification of candidate substrates. *Am J Physiol* **270**, F319-325.

Kankaanpaa, H. T., Holliday, J., Schroder, H., Goddard, T. J., von Fister, R., and Carmichael, W. W. (2005). Cyanobacteria and prawn farming in northern New South Wales, Australia--a case study on cyanobacteria diversity and hepatotoxin bioaccumulation. *Toxicol Appl Pharmacol* **203**, 243-256.

Kenyon, C. N., Rippka, R., and Stanier, R. Y. (1972). Fatty acid composition and physiological properties of some filamentous blue-green algae. *Arch Mikrobiol* **83**, 216-236.

Khan, S. A., Wickstrom, M., Haschek, W., Schaeffer, S., Ghosh, S., and Beasley, V. (1996). Microcystin-LR and kinetics of cytoskeletal reorganization in hepatocytes, kidney cells, and fibroblasts. *Natural Toxins* **4**, 206-214.

Kim, D., Su, J., and Cotman, C. W. (1999). Sequence of neurodegeneration and accumulation of phosphorylated tau in cultured neurons after okadaic acid treatment. *Brain Res* **839**, 253-262.

Kis, B., Chen, L., Ueta, Y., and Busija, D. W. (2006). Autocrine peptide mediators of cerebral endothelial cells and their role in the regulation of blood-brain barrier. *Peptides* **27**, 211-222.

Komatsu, M., Furukawa, T., Ikeda, R., Takumi, S., Nong, Q., Aoyama, K., Akiyama, S., Keppler, D., and Takeuchi, T. (2007). Involvement of mitogen-activated protein kinase signaling pathways in microcystin-LR-induced apoptosis after its selective uptake mediated by OATP1B1 and OATP1B3. *Toxicol Sci* **97**, 407-416.

König, J., Cui, Y., Nies, A. T., and Keppler, D. (2000). A novel human organic anion transporting polypeptide localized to the basolateral hepatocyte membrane. *Am J Physiol Gastrointest Liver Physiol* **278**, G156-164.

König, J., Cui, Y., Nies, A. T., and Keppler, D. (2000). Localization and genomic organization of a new hepatocellular organic anion transporting polypeptide. *The journal of biological chemistry* **275**, 23161-23168.

Konig, J., Seithel, A., Gradhand, U., and Fromm, M. F. (2006). Pharmacogenomics of human OATP transporters. *Naunyn Schmiedebergs Arch Pharmacol* **372**, 432-443.

Kotak, B. G., Kenefick, S. L., Fritz, D. L., Rousseaux, C. G., Prepas, E. E., and Hrudey, S. E. (1993). Occurrence and toxicolological evaluation of cyanobacterial toxins in Alberta lakes and farm dugouts. *Water Research* **27**, 495-506.

Kotak, B. G., Hrudey, S. E., Kenefick, S. L., and Prepas, E. E. (1993a). Toxicity of cyanobacterial blooms in Alberta lakes. Annual Aquatic Toxicity Workshop, pp. 172-179.

Krakstad, C., Herfindal, L., Gjertsen, B. T., Boe, R., Vintermyr, O. K., Fladmark, K. E., and Doskeland, S. O. (2006). CaM-kinaseII-dependent commitment to microcystin-induced apoptosis is coupled to cell budding, but not to shrinkage or chromatin hypercondensation. *Cell Death Differ* **13**, 1191-1202.

Krienitz, L., Ballot, A., Wiegand, C., Kotut, K., Codd, G. A., and Pflugmacher, S. (2002). Cyanotoxin-producing bloom of *Anabaena flos-aquae*, *Anabaena discoidea* and *Microcystis aeruginosa* (Cyanobacteria) in Nyanza Gulf of Lake Victoria, Kenya. *Journal of Applied Botany* [print] **76**, 179-183.

Krienitz, L., Ballot, A., Kotut, K., Wiegand, C., Puetz, S., Metcalf, J. S., Codd, G. A., and Pflugmacher, S. (2003). Contribution of hot spring cyanobacteria to the mysterious deaths of Lesser Flamingos at Lake Bogoria, Kenya. *FEMS Microbiology Ecology* **43**, 141-148.

Krishnamurthy, T., Szafraniec, L., Hunt, D. F., Shabanowitz, J., Yates III, J. R., Hauer, C. R., Carmichael, W. W., Skulberg, O., Codd, G. A., and Missler, S. (1989). Structural characterization of toxic cyclic peptides from blue-green algae by tandem mass spectrometry. *Proceedings of the National Academy of Sciences of the USA* **86**, 770-774.

Kuiper-Goodman, T., Falconer, I. R., and Fitzgerald, D. J. (1999). Human Health Aspects. In *Toxic Cyanobacteria in Water: A Guide to their Public Health Consequences, Monitoring and Management* (I. Chorus, and J. Bartram, Eds.), pp. 114-153. E & FN Spon, London.

Kullak-Ublick, G.-A., Hagenbuch, B., Stieger, B., Wolkoff, A. W., and Meier, P. J. (1994). Functional characterization of the basolateral rat liver organic anion transporting polypeptide. *Hepatology* **20**, 411-416.

Kullak-Ublick, G. A., Hagenbuch, B., Stieger, B., Schteingart, C. D., Hofmann, A. F., Wolkoff, A. W., and Meier, P. J. (1995). Molecular and functional characterization of an organic anion transporting polypeptide cloned from human liver. *Gastroenterology* **109**, 1274-1282.

Kullak-Ublick, G. A., Stieger, B., and Meier, P. J. (2004). Enterohepatic bile salt transporters in normal physiology and liver disease. *Gastroenterology* **126**, 322-342.

Kusumi, T., Ooi, T., Watanabe, M. M., Takahashi, H., and Kakisawa, H. (1987). Cyanoviridin RR, a toxin from the cyanobacterium (blue-green alga) *Microcystis viridis*. *Tetrahedron Letters* **28**, 4695-4698.

Laemmli, U. (1970). Cleavage of structural proteins during assembly of the head of bacteriophage T4. *Nature* **227**, 680-685.

7. References

Lanaras, T., Tsitsamis, S., Chlichlia, C., and Cook, C. M. (1989). Toxic cyanobacteria in Greek freshwaters. *Journal of Applied Phycology* **1**, 67-73.

Lanaras, T., and Cook, C. M. (1994). Toxin extraction from an Anabaenopsis milleri--dominated bloom. *Sci Total Environ* **142**, 163-169.

Lankoff, A., Banasik, A., Obe, G., Deperas, M., Kuzminski, K., Tarczynska, M., Jurczak, T., and Wojcik, A. (2003). Effect of microcystin-LR and cyanobacterial extract from Polish reservoir of drinking water on cell cycle progression, mitotic spindle, and apoptosis in CHO-K1 cells. *Toxicol Appl Pharmacol* **189**, 204-213.

Lankoff, A., Krzowski, L., Glab, J., Banasik, A., Lisowska, H., Kuszewski, T., Gozdz, S., and Wojcik, A. (2004). DNA damage and repair in human peripheral blood lymphocytes following treatment with microcystin-LR. *Mutation Research* **559**, 131-142.

Lankoff, A., Carmichael, W. W., Grasman, K. A., and Yuan, M. (2004a). The uptake kinetics and immunotoxic effects of microcystin-LR in human and chicken peripheral blood lymphocytes in vitro. *Toxicology* **204**, 23-40.

Lawrence, J. F., Niedzwiadek, B., Menard, C., Lau, B. P., Lewis, D., Kuiper-Goodman, T., Carbone, S., and Holmes, C. (2001a). Comparison of liquid chromatography/mass spectrometry, ELISA, and phosphatase assay for the determination of microcystins in blue-green algae products. *Journal of the AOAC International* **84**, 1035-1044.

Lee, W., Glaeser, H., Smith, L. H., Roberts, R. L., Moeckel, G. W., Gervasini, G., Leake, B. F., and Kim, R. B. (2005). Polymorphisms in human organic anion-transporting polypeptide 1A2 (OATP1A2): implications for altered drug disposition and central nervous system drug entry. *J Biol Chem* **280**, 9610-9617.

Leist, M., and Jaattela, M. (2001). Four deaths and a funeral: from caspases to alternative mechanisms. *Nat Rev Mol Cell Biol* **2**, 589-598.

Letschert, K., Komatsu, M., Hummel-Eisenbeiss, J., and Keppler, D. (2005). Vectorial transport of the peptide CCK-8 by double-transfected MDCKII cells stably expressing the organic anion transporter OATP1B3 (OATP8) and the export pump ABCC2. *J Pharmacol Exp Ther* **313**, 549-556.

Lindholm, T., and Meriluoto, J. A. O. (1991). Recurrent depth maxima of the hepatotoxic Cyanobacterium *Oscillatoria agardhii*. *Canadian Journal of Fisheries and Aquatic Sciences* **48**, 1629-1634.

Lindholm, T., Ohman, P., Kurki-Helasmo, K., Kincaid, B., and Meriluoto, J. (1999). Toxic algae and fish mortality in a brackish-water lake in Aland, SW Finland. *Hydrobiologia* **397**, 109-120.

Lippy, E. C., and Erb, J. (1976). Gastrointestinal illness at Sewickley. *Journal of the American Water Works Association* **68**, 606-610.

Liu, F., Grundke-Iqbal, I., Iqbal, K., and Gong, C. X. (2005). Contributions of protein phosphatases PP1, PP2A, PP2B and PP5 to the regulation of tau phosphorylation. *Eur J Neurosci* **22**, 1942-1950.

Llewellyn, L. E. (2006). Saxitoxin, a toxic marine natural product that targets a multitude of receptors. *Nat Prod Rep* **23**, 200-222.

Lu, H., Choudhuri, S., Ogura, K., Csanaky, I. L., Lei, X., Cheng, X., Song, P. Z., and Klaassen, C. D. (2008). Characterization of organic anion transporting polypeptide 1b2-null mice: essential role in hepatic uptake/toxicity of phalloidin and microcystin-LR. *Toxicol Sci* **103**, 35-45.

MacKintosh, C., Beattie, K. A., Klumpp, S., Cohen, P., and Codd, G. A. (1990). Cyanobacterial microcystin-LR is a potent and specific inhibitor of protein phosphatases 1 and 2A from both mammals and higher plants. *FEBS Letters* **264**, 187.

MacKintosh, C. (1993). Assay and purification of protein (serine/threonine) phosphatases. In *Protein Phosphorylation: A Practical Approach* (D. G. Hardie, Ed.), pp. 197-229. Oxford University Press, Oxford.

MacKintosh, C., and MacKintosh, R. W. (1994). Inhibitors of protein kinases and phosphatases. *Trends in Biochemical Science* **19**, 444-448.

MacKintosh, R. W., Dalby, K. N., Campbell, D. G., Cohen, P. T. W., Cohen, P., and MacKintosh, C. (1995). The cyanobacterial toxin microcystin binds covalently to cysteine-273 on protein phosphatase 1. *FEBS Letters* **371**, 236.

Magalhaes, V. F., Soares, R. M., and Azevedo, S. M. (2001). Microcystin contamination in fish from the Jacarepagua Lagoon (Rio de Janeiro, Brazil): ecological implication and human health risk. *Toxicon* **39**, 1077-1085.

Magalhaes, V. F., Marinho, M. M., Domingos, P., Oliveira, A. C., Costa, S. M., Azevedo, L. O., and Azevedo, S. M. (2003). Microcystins (cyanobacteria hepatotoxins) bioaccumulation in fish and crustaceans from Sepetiba Bay (Brasil, RJ). *Toxicon* **42**, 289-295.

Maidana, M., Carlis, V., Galhardi, F. G., Yunes, J. S., Geracitano, L. A., Monserrat, J. M., and Barros, D. M. (2006). Effects of microcystins over short- and long-term memory and oxidative stress generation in hippocampus of rats. *Chem Biol Interact* **159**, 223-234.

Mankiewicz, J., Tarczynska, M., Fladmark, K. E., Doskeland, S. O., Walter, Z., and Zalewski, M. (2001). Apoptotic effect of cyanobacterial extract on rat hepatocytes and human lymphocytes. *Environ Toxicol* **16**, 225-233.

Maruyama, T., Kato, K., Yokoyama, A., Tanaka, T., Hiraishi, A., and Park, H. D. (2003). Dynamics of microcystin-degrading bacteria in mucilage of microcystis. *Microbial Ecology* **46**, 279-288.

Matsunaga, H., Harada, K.-I., Senma, M., Ito, Y., Yasuda, N., Ushida, S., and Kimura, Y. (1999). Possible cause uf unnatural mass death of wild birds in a pond in Nishinomiya, Japan: sudden appearance of toxic cyanobacteria. *Natural Toxins* **7**, 81-84.

Maynes, J. T., Luu, H. A., Cherney, M. M., Andersen, R. J., Williams, D., Holmes, C. F., and James, M. N. (2006). Crystal structures of protein phosphatase-1 bound to motuporin and dihydromicrocystin-LA: elucidation of the mechanism of enzyme inhibition by cyanobacterial toxins. *J Mol Biol* **356**, 111-120.

7. References

McCluskey, A., Ackland, S. P., Gardiner, E., Walkom, C. C., and Sakoff, J. A. (2001). The inhibition of protein phosphatases 1 and 2A: a new target for rational anti-cancer drug design? *Anticancer Drug Des* **16**, 291-303.

McCluskey, A., Sim, A. T., and Sakoff, J. A. (2002). Serine-threonine protein phosphatase inhibitors: development of potential therapeutic strategies. *Journal of Medicinal Chemistry* **45**, 1151-1175.

Meier-Abt, F., Hammann-Hanni, A., Stieger, B., Ballatori, N., and Boyer, J. L. (2007). The organic anion transport polypeptide 1d1 (Oatp1d1) mediates hepatocellular uptake of phalloidin and microcystin into skate liver. *Toxicol Appl Pharmacol* **218**, 274-279.

Meier, P. J., and Stieger, B. (2002). Bile salt transporters. *Annual Review of Physiology* **64**, 635-661.

Meriluoto, J. A., Nygard, S. E., Dahlem, A. M., and Eriksson, J. E. (1990). Synthesis, organotropism and hepatocellular uptake of two tritium-labeled epimers of dihydromicrocystin-LR, a cyanobacterial peptide toxin analog. *Toxicon* **28**, 1439-1446.

Meriluoto, J. A., and Spoof, L. E. (2008). Cyanotoxins: sampling, sample processing and toxin uptake. *Adv Exp Med Biol* **619**, 483-499.

Metcalf, J. S., and Codd, G. A. (2004). Cyanobacterial toxins in the water environment, pp. 36. Foundation for Water Research, Marlow.

Meyer Zu Schwabedissen, H. E., Ware, J. A., Tirona, R. G., and Kim, R. B. (2009). Identification, Expression, and Functional Characterization of Full-Length and Splice Variants of Murine Organic Anion Transporting Polypeptide 1b2. *Mol Pharm.*

Mez, K., Hanselmann, K., Naegeli, H., and Preisig, H. R. (1996). Protein phosphatase-inhibiting activity in cyanobacteria from alpine lakes in Switzerland. *Phycologia* **35**, 133-139.

Mez, K., Beattie, K., Codd, G., Hanselmann, K., Hauser, B., Naegeli, H., and Preisig, H. (1997). Identification of a microcystin in benthic cyanobacteria linked to cattle deaths on alpine pastures in Switzerland. *European Journal of Phycology* **32**, 111-117.

Mez, K. (1997a). Toxin production by benthic cyanobacteria in oligotrophic Alpine lakes. In *Philosophische Fakultät II*, pp. 75. Universität Zürich, Zürich.

Mikhailov, A., Harmala-Brasken, A.-S., Polosukhina, E., Hanski, A., Wahlsten, M., Sivonen, K., and Eriksson, J. E. (2001). Production and specificity of monoclonal antibodies against nodularin conjugated through N-methyldehydrobutyrine. *Toxicon* **39**, 1453-1459.

Mikhailov, A., Harmala-Brasken, A. S., Hellman, J., Meriluoto, J., and Eriksson, J. E. (2003). Identification of ATP-synthase as a novel intracellular target for microcystin-LR. *Chem Biol Interact* **142**, 223-237.

Mikkaichi, T., Suzuki, T., Tanemoto, M., Ito, S., and Abe, T. (2004). The organic anion transporter (OATP) family. *Drug Metab Pharmacokinet* **19**, 171-179.

Miller, S. R., Castenholz, R. W., and Pedersen, D. (2007). Phylogeography of the Thermophilic Cyanobacterium Mastigocladus laminosus. *Appl. Environ. Microbiol.* **73**, 4751-4759.

Milutinovic, A., Sedmak, B., Horvat-Znidarsic, I., and Suput, D. (2002). Renal injuries induced by chronic intoxication with microcystins. *Cellular and Molecular Biology Letters* **7**, 139-141.

Milutinovic, A., Zivin, M., Zorc-Pleskovic, R., Sedmak, B., and Suput, D. (2003). Nephrotoxic effects of chronic administration of microcystins -LR and -YR. *Toxicon* **42**, 281-288.

Miura, G. A., Robinson, N. A., Lawrence, W. B., and Pace, J. G. (1991). Hepatotoxicity of microcystin-LR in fed and fasted rats. *Toxicon* **29**, 337-346.

Monks, N. R., Liu, S., Xu, Y., Yu, H., Bendelow, A. S., and Moscow, J. A. (2007). Potent cytotoxicity of the phosphatase inhibitor microcystin LR and microcystin analogues in OATP1B1- and OATP1B3-expressing HeLa cells. *Mol Cancer Ther* **6**, 587-598.

Mosmann, T. (1983). Rapid colorimetric assay for cellular growth and survival: application to proliferation and cytotoxicity assays. *Journal of immunological methods* **65**, 55-63.

Moura, S., Ultramari Mde, A., de Paula, D. M., Yonamine, M., and Pinto, E. (2009). 1H NMR determination of beta-N-methylamino-L-alanine (L-BMAA) in environmental and biological samples. *Toxicon* **53**, 578-583.

Mühlhardt, C. (2002). Der Experimentator: Molekularbiologie, 3 Auflage, Spektrum, Heidelberg, Germany

Mur, L. R., Skulberg, O. M., and Utkilen, H. (1999). Cyanobacteria in the environment. In *Toxic Cyanobacteria in Water: A Guide to their Public Health Consequences, Monitoring and Management* (I. Chorus, and J. Bartram, Eds.), pp. 15-40. E & FN Spon, London.

Murch, S. J., Cox, P. A., Banack, S. A., Steele, J. C., and Sacks, O. W. (2004). Occurrence of beta-methylamino-l-alanine (BMAA) in ALS/PDC patients from Guam. *Acta Neurol Scand* **110**, 267-269.

Myers, J. L., and Richardson, L. L. (2009). Adaptation of cyanobacteria to the sulfide-rich microenvironment of black band disease of coral. *FEMS Microbiol Ecol* **67**, 242-251.

Naegeli, H., Sahin, A., Braun, U., Hauser, B., Mez, K., Hanselmann, K., Preisig, H. R., Bivetti, A., and Eitel, J. (1997). Sudden deaths of cattle on Alpine pastures in South-Eastern Switzerland. [German]. *Schweizer Archiv Fuer Tierheilkunde* **139**, 201-209.

Namikoshi, M., Choi, B. W., Sun, F., Rinehart, K. L., Evans, W. R., and Carmichael, W. W. (1993). Chemical characterization and toxicity of dihydro derivatives of nodularin and microcystin-LR, potent cyanobacterial cyclic peptide hepatotoxins. *Chemical Research and Toxicology* **6**, 151-158.

7. References

Naud, J., Michaud, J., Boisvert, C., Desbiens, K., Leblond, F. A., Mitchell, A., Jones, C., Bonnardeaux, A., and Pichette, V. (2007). Down-regulation of intestinal drug transporters in chronic renal failure in rats. *J Pharmacol Exp Ther* **320**, 978-985.

Negri, A. P., and Jones, G. J. (1995). Bioaccumulation of paralytic shellfish poisoning (PSP) toxins from the cyanobacterium Anabaena circinalis by the freshwater mussel *Alathyria condola*. *Toxicon* **33**, 667-678.

Negri, A. P., Jones, G. J., Blackburn, S. I., Oshima, Y., and Onodera, H. (1997). Effect of culture and bloom development and of sample storage on paralytic shellfish poisons in the cyanobacterium Anabaena circinalis. *Journal of Phycology* **33**, 26-35.

Neilan, B. A., Burns, B. P., Relman, D. A., and Lowe, D. R. (2002). Molecular identification of cyanobacteria associated with stromatolites from distinct geographical locations. *Astrobiology* **2**, 271-280.

Nies, A. T. (2007). The role of membrane transporters in drug delivery to brain tumors. *Cancer Lett* **254**, 11-29.

Nishiwaki-Matsushima, R., Ohta, T., Nishiwaki, S., Suganuma, M., Kohyama, K., Ishikawa, T., Carmichael, W. W., and Fujiki, H. (1992). Liver tumor promotion by the cyanobacterial cyclic peptide toxin microcystin-LR. *J Cancer Res Clin Oncol* **118**, 420-424.

Nishiwaki, R., Ohta, T., Sueoka, E., Suganuma, M., Harada, K., Watanabe, M. F., and Fujiki, H. (1994). Two significant aspects of microcystin-LR: specific binding and liver specificity. *Cancer Lett* **83**, 283-289.

Nobre, A. C., Jorge, M. C., Menezes, D. B., Fonteles, M. C., and Monteiro, H. S. (1999). Effects of microcystin-LR in isolated perfused rat kidney. *Braz J Med Biol Res* **32**, 985-988.

Nobre, A. C., Coelho, G. R., Coutinho, M. C., Silva, M. M., Angelim, E. V., Menezes, D. B., Fonteles, M. C., and Monteiro, H. S. (2001). The role of phospholipase A(2) and cyclooxygenase in renal toxicity induced by microcystin-LR. *Toxicon* **39**, 721-724.

Nobre, A. C., Nunes-Monteiro, S. M., Monteiro, M. C., Martins, A. M., Havt, A., Barbosa, P. S., Lima, A. A., and Monteiro, H. S. (2004). Microcystin-LR promote intestinal secretion of water and electrolytes in rats. *Toxicon* **44**, 555-559.

Oberholster, P. J., Myburgh, J. G., Govender, D., Bengis, R., and Botha, A. M. (2009). Identification of toxigenic Microcystis strains after incidents of wild animal mortalities in the Kruger National Park, South Africa. *Ecotoxicol Environ Saf* **72**, 1177-1182.

Ohta, T., Nishiwaki, R., Yatsunami, J., Komori, A., Suganuma, M., and Fujiki, H. (1992). Hyperphosphorylation of cytokeratins 8 and 18 by microcystin-LR, a new liver tumor promoter, in primary cultured rat hepatocytes. *Carcinogenesis* **13**, 2443-2447.

Ohta, T., Sueoka, E., Iida, N., Komori, A., Suganuma, M., Nishiwaki, R., Tatematsu, M., Kim, S. J., Carmichael, W. W., and Fujiki, H. (1994). Nodularin, a potent inhibitor of protein phosphatases 1 and 2A, is a new environmental carcinogen in male F344 rat liver. *Cancer Research* **54**, 6402-6406.

Ohtsuki, S., Takizawa, T., Takanaga, H., Hori, S., Hosoya, K., and Terasaki, T. (2004). Localization of organic anion transporting polypeptide 3 (oatp3) in mouse brain parenchymal and capillary endothelial cells. *J Neurochem* **90**, 743-749.

Oliver, R. L. (1994). Floating and sinking in gas-vacuolate cyanobacteria. *Journal of Phycology* **30**, 161-173.

Oliver, R. L., and Walsby, A. E. (1984). Direct evidence for the role of light-mediated gas-vesicle collapse in the buoyancy regulation of Anabaena flos-aquae (cyanobacteria). *Limnology and Oceanography.* **29**, 879-886.

Onodera, H., Oshima, Y., Henriksen, P., and Yasumoto, T. (1997). Confirmation of anatoxin-a(s), in the cyanobacterium Anabaena lemmermannii, as the cause of bird kills in Danish lakes. *Toxicon* **35**, 1645-1648.

Oren, A., and Gunde-Cimerman, N. (2007). Mycosporines and mycosporine-like amino acids: UV protectants or multipurpose secondary metabolites? *FEMS Microbiol Lett* **269**, 1-10.

Osswald, J., Rellan, S., Gago, A., and Vasconcelos, V. (2007). Toxicology and detection methods of the alkaloid neurotoxin produced by cyanobacteria, anatoxin-a. *Environ Int* **33**, 1070-1089.

Paerl, H. W., Steppe, T. F., and Reid, R. P. (2001). Bacterially mediated precipitation in marine stromatolites. *Environ Microbiol* **3**, 123-130.

Paerl, H. W., Steppe, T. F., Buchan, K. C., and Potts, M. (2003). Hypersaline cyanobacterial mats as indicators of elevated tropical hurricane activity and associated climate change. *Ambio* **32**, 87-90.

Paerl, H. W., and Huisman, J. (2008). Climate. Blooms like it hot. *Science* **320**, 57-58.

Paerl, H. (2008a). Nutrient and other environmental controls of harmful cyanobacterial blooms along the freshwater-marine continuum. *Adv Exp Med Biol* **619**, 217-237.

Page, R. D. (1996). TreeView: an application to display phylogenetic trees on personal computers. *Comput Appl Biosci* **12**, 357-358.

Park, H. D., Sasaki, Y., Maruyama, T., Yanagisawa, E., Hiraishi, A., and Kato, K. (2001). Degradation of the cyanobacterial hepatotoxin microcystin by a new bacterium isolated from a hypertrophic lake. *Environmental Toxicology* **16**, 337-343.

Park, H., Namikoshi, M., Brittian, S., Carmichael, W., and Murphy, T. (2001a). [D-Leu^1] microcystin-LR, a new microcystin isolated from waterbloom in a Canadian prairie lake. *Toxicon* **39**, 855-862.

Paul, V. J. (2008). Global warming and cyanobacterial harmful algal blooms. *Adv Exp Med Biol* **619**, 239-257.

Pei, J.-J., Gong, C.-X., An, W.-L., Winblad, B., Cowburn, R. F., Grundke-Iqbal, I., and Iqbal, K. (2003). Okadaic-Acid-Induced Inhibition of Protein Phosphatase 2A Produces Activation of Mitogen-Activated Protein Kinases ERK1/2, MEK1/2, and p70 S6, Similar to That in Alzheimer's Disease. *Am J Pathol* **163**, 845-858.

7. References

Pelroy, R. A., Rippka, R., and Stanier, R. Y. (1972). Metabolism of glucose by unicellular blue-green algae. *Arch Mikrobiol* **87**, 303-322.

Perry, T. L., Bergeron, C., Biro, A. J., and Hansen, S. (1989). Beta-N-methylamino-L-alanine. Chronic oral administration is not neurotoxic to mice. *J Neurol Sci* **94**, 173-180.

Persson, P.-E., Sivonen, K., Keto, J., Kononen, K., Niemi, M., and Viljamaa, H. (1984). Potentially toxic blue-green algae (Cyanobacteria) in Finnish natural waters. *Aqua Fennica* **14**, 147-154.

Pilotto, L., Douglas, R., Burch, M., Cameron, S., Beers, M., Rouch, G., Robinson, P., Kirk, M., Cowie, C., Hardiman, S., Moore, C., and Attewell, R. (1997). Health effects of exposure to cyanobacteria (blue-green algae) during recreational water-related activities. *Australian and New Zealand Journal of Public Health* **21**, 562-566.

Pizzagalli, F., Hagenbuch, B., Stieger, B., Klenk, U., Folkers, G., and Meier, P. J. (2002). Identification of a novel human organic anion transporting polypeptide as a high affinity thyroxine transporter. *Mol Endocrinol* **16**, 2283-2296.

Post, A. F., and Arieli, B. (1997). Photosynthesis of Prochlorothrix hollandica under Sulfide-Rich Anoxic Conditions. *Appl Environ Microbiol* **63**, 3507-3511.

Pouria, S., de Andrade, A., Barbosa, J., Cavalcanti, R. L., Barreto, V. T., Ward, C. J., Preiser, W., Poon, G. K., Neild, G. H., and Codd, G. A. (1998). Fatal microcystin intoxication in haemodialysis unit in Caruaru, Brazil. *Lancet* **352**, 21-26.

Rametti, A., Esclaire, F., Yardin, C., and Terro, F. (2004). Linking alterations in tau phosphorylation and cleavage during neuronal apoptosis. *J Biol Chem* **279**, 54518-54528.

Rapala, J., Robertson, A., Negri, A. P., Berg, K. A., Tuomi, P., Lyra, C., Lahti, K., Hoppu, K., and Lepistö, L. (2004). First report of saxitoxin in an associated adverses health effects in lakes of Finland. ICTC 6th.

Rapala, J., Robertson, A., Negri, A. P., Berg, K. A., Tuomi, P., Lyra, C., Erkomaa, K., Lahti, K., Hoppu, K., and Lepisto, L. (2005). First report of saxitoxin in Finnish lakes and possible associated effects on human health. *Environ Toxicol* **20**, 331-340.

Rapala, J., Berg, K. A., Lyra, C., Niemi, R. M., Manz, W., Suomalainen, S., Paulin, L., and Lahti, K. (2005a). Paucibacter toxinivorans gen. nov., sp. nov., a bacterium that degrades cyclic cyanobacterial hepatotoxins microcystins and nodularin. *Int J Syst Evol Microbiol* **55**, 1563-1568.

Reid, R. P., Visscher, P. T., Decho, A. W., Stolz, J. F., Bebout, B. M., Dupraz, C., Macintyre, I. G., Paerl, H. W., Pinckney, J. L., Prufert-Bebout, L., Steppe, T. F., and DesMarais, D. J. (2000). The role of microbes in accretion, lamination and early lithification of modern marine stromatolites. *Nature* **406**, 989-992.

Reynolds, C. S., Jaworski, G. H. M., Cmiech, H. A., and Leedale, G. F. (1981). On the annual cycle of the blue-green alga *Microcystis aeruginosa* Kütz Emend. Elenkin. *Proceedings and Philosophical Transactions of the Royal Society of London, B. Biological Sciences* **293**, 419.

Rippka, R., Deruelles, J., Waterbury, J. B., Herdman, M., and Stanier, R. Y. (1979). Generic assignments, strain histories and properties of pure cultures of cyanobacteria. *Journal of General Microbiology* **111**, 1-61.

Robinson, N. A., Miura, G. A., Matson, C. F., Dinterman, R. E., and Pace, J. G. (1989). Characterization of chemically tritiated microcystin-LR and its distribution in mice. *Toxicon* **27**, 1035-1042.

Rodrigue, D. C., Etzel, R. A., Hall, S., de Porras, E., Velasquez, O. H., Tauxe, R. V., Kilbourne, E. M., and Blake, P. A. (1990). Lethal paralytic shellfish poisoning in Guatemala. *Am J Trop Med Hyg* **42**, 267-271.

Rossini, G. P., Pinna, C., and Malaguti, C. (1999). Different sensitivities of p42 mitogen-activated protein kinase to phorbol ester and okadaic acid tumor promoters among cell types. *Biochem Pharmacol* **58**, 279-284.

Rossini, G. P., Sgarbi, N., and Malaguti, C. (2001). The toxic responses induced by okadaic acid involve processing of multiple caspase forms. *Toxicon* **39**, 563-770.

Rozen, S., Skaletzky H. J. (2000). Primer3 on the WWW for general users and biologist programmers. In: Krawetz, S., Misener, S. (Eds), Bioinformatic Methods and Protocols: Methods in Molecular Biology. Human Press, Totowa, NJ, pp. 365-386.

Runnegar, M. T. C., Gerdes, R. G., and Falconer, I. R. (1991). The uptake of the cyanobacterial hepatotoxin microcystin by isolated rat hepatocytes. *Toxicon* **29**, 43.

Runnegar, M., Berndt, N., and Kaplowitz, N. (1995). Microcystin uptake and inhibition of protein phosphatases: effects of chemoprotectants and self-inhibition in relation to known hepatic transporters. *Toxicol Appl Pharmacol* **134**, 264-272.

Runnegar, M., Berndt, N., Kong, S. M., Lee, E. Y., and Zhang, L. (1995a). In vivo and in vitro binding of microcystin to protein phosphatases 1 and 2A. *Biochem Biophys Res Commun* **216**, 162-169.

Sahara, M. M., Boyoung Lee, Jung-Mi Park, Sarita Lagalwar, Lester I. Binder, Akihiko Takashima, (2008). Active c-jun N-terminal kinase induces caspase cleavage of tau and additional phosphorylation by GSK-3β; is required for tau aggregation. *European Journal Of Neuroscience* **27**, 2897-2906.

Sai, Y., Kaneko, Y., Ito, S., Mitsuoka, K., Kato, Y., Tamai, I., Artursson, P., and Tsuji, A. (2006). Predominant contribution of organic anion transporting polypeptide OATP-B (OATP2B1) to apical uptake of estrone-3-sulfate by human intestinal Caco-2 cells. *Drug Metab Dispos* **34**, 1423-1431.

Saker, M. L., and Eaglesham, G. K. (1999). The accumulation of cylindrospermopsin from the cyanobacterium *Cylindrospermopsis raciborskii* in tissues of the Redclaw crayfish *Cherax quadricarinatus*. *Toxicon* **37**, 1065-1077.

Sano, T., Takagi, H., Sadakane, K., Ichinose, T., Kawazato, H., and Kaya, K. (2004). Cacionogenic effects of microcystin-LR and Dhb-microcystin-LR on mice liver. ICTC 6th.

Sato, K., Sugawara, J., Sato, T., Mizutamari, H., Suzuki, T., Ito, A., Mikkaichi, T., Onogawa, T., Tanemoto, M., Unno, M., Abe, T., and Okamura, K. (2003). Expression of organic anion transporting polypeptide E (OATP-E) in human placenta. *Placenta* **24**, 144-148.

7. References

Schaeffer, D. J., Malpas, P. B., and Barton, L. L. (1999). Risk Assessment of Microcystin in Dietary Aphanizomenon flos-aquae. *Ecotoxicology and Environmental Safety* **44**, 73.
Schopf, J. W., and Packer, B. M. (1987). Early Archean (3.3-billion to 3.5-billion-year-old) microfossils from Warrawoona Group, Australia. *Science* **237**, 70-73.
Sergeev, V. N., Gerasimenko, L. M., and Zavarzin, G. A. (2002). [Proterozoic history and present state of cyanobacteria]. *Mikrobiologiia* **71**, 725-740.
Sevrin-Reyssac, J., and Pletikosic, M. (1990). Cyanobacteria in fish ponds. *Aquaculture* **88**, 1-20.
Sivonen, K., Niemelä, S. I., Niemi, R. M., Lepistö, L., Luoma, T. H., and Räsänen, L. A. (1990). Toxic cyanobacteria (blue-green) algae in Finnish fresh and coastal waters. *Hydrobiologia* **190**, 267-275.
Sivonen, K. (1990a). Effects of light, temperature, nitrate, orthophosphate, and bacteria on growth of and hepatotoxin production by *Oscillatoria agardhii* strains. *Applied Environmental Microbiology* **56**, 2658-2666.
Sivonen, K., Namikoshi, M., Evans, W. R., Carmichael, W. W., Sun, F., Rouhiainen, L., Luukkainen, R., and Rinehart, K. L. (1992). Isolation and characterization of a variety of microcystins from seven strains of the cyanobacterial genus *Anabaena*. *Applied Environmental Microbiology* **58**, 2495-2500.
Sivonen, K., and Jones, G. (1999). Cyanobacterial toxins. In *Toxic Cyanobacteria in Water: A Guide to their Public Health Consequences, Monitoring and Management* (I. Chorus, and J. Bartram, Eds.), pp. 41-111. E & FN Spon, London.
Skaper, S. D., Buriani, A., Dal Toso, R., Petrelli, L., Romanello, S., Facci, L., and Leon, A. (1996). The ALIAmide palmitoylethanolamide and cannabinoids, but not anandamide, are protective in a delayed postglutamate paradigm of excitotoxic death in cerebellar granule neurons. *Proc Natl Acad Sci U S A* **93**, 3984-3989.
Skulberg, O. M., Carmichael, W. W., Andersen, R. A., Matsunaga, S., Moore, R. E., and Skulberg, R. (1992). Investigations of a Neurotoxic Oscillatorialean Strain Cyanophyceae and Its Toxin Isolation and Characterization of Homoanatoxin A. *Environmental Toxicology & Chemistry* **11**, 321-329.
Smith, Q. R., Nagura, H., Takada, Y., and Duncan, M. W. (1992). Facilitated transport of the neurotoxin, beta-N-methylamino-L-alanine, across the blood-brain barrier. *J Neurochem* **58**, 1330-1337.
Soares, R. M., Yuan, M., Servaites, J. C., Delgado, A., Magalhaes, V. F., Hilborn, E. D., Carmichael, W. W., and Azevedo, S. M. (2006). Sublethal exposure from microcystins to renal insufficiency patients in Rio de Janeiro, Brazil. *Environ Toxicol* **21**, 95-103.
Solter, P. F., Wollenberg, G. K., Huang, X., Chu, F. S., and Runnegar, M. T. (1998). Prolonged sublethal exposure to the protein phosphatase inhibitor microcystin-LR results in multiple dose-dependent hepatotoxic effects. *Toxicological Sciences* **44**, 87-96.
Spencer, P. S., Nunn, P. B., Hugon, J., Ludolph, A. C., Ross, S. M., Roy, D. N., and Robertson, R. C. (1987). Guam amyotrophic lateral sclerosis-parkinsonism-dementia linked to a plant excitant neurotoxin. *Science* **237**, 517-522.
Spoof, L. (2005). Microcystins and nodularins. In *TOXIC: cyanobacterial monitoring and cyanotoxin analysis* (J. Meriluoto, and G. A. Codd, Eds.), pp. 15-39. Abo Akademi University Press.
Sroga, G. E. (1997). Regulation of nitrogen fixation by different nitrogen sources in the filamentous non-heterocystous cyanobacterium Microcoleus sp. *FEMS Microbiol Lett* **153**, 11-15.
Stanier, R. Y., and Cohen-Bazire, G. (1977). Phototrophic prokaryotes: the cyanobacteria. *Annu Rev Microbiol* **31**, 225-274.
Stewart, W. D. (1973). Nitrogen fixation by photosynthetic microorganisms. *Annu Rev Microbiol* **27**, 283-316.
Stewart, W. D., Rowell, P., and Tel-or, E. (1975). Nitrogen fixation and the heterocyst in blue-green algae. *Biochem Soc Trans* **3**, 357-361.
Stewart, I., Seawright, A. A., and Shaw, G. R. (2008). Cyanobacterial poisoning in livestock, wild mammals and birds--an overview. *Adv Exp Med Biol* **619**, 613-637.
Sugiyama, D., Kusuhara, H., Taniguchi, H., Ishikawa, S., Nozaki, Y., Aburatani, H., and Sugiyama, Y. (2003). Functional characterization of rat brain-specific organic anion transporter (Oatp14) at the blood-brain barrier: high affinity transporter for thyroxine. *J Biol Chem* **278**, 43489-43495.
Summers, M. L., Wallis, J. G., Campbell, E. L., and Meeks, J. C. (1995). Genetic evidence of a major role for glucose-6-phosphate dehydrogenase in nitrogen fixation and dark growth of the cyanobacterium Nostoc sp. strain ATCC 29133. *J Bacteriol* **177**, 6184-6194.
Tamagnini, P., Leitao, E., Oliveira, P., Ferreira, D., Pinto, F., Harris, D. J., Heidorn, T., and Lindblad, P. (2007). Cyanobacterial hydrogenases: diversity, regulation and applications. *FEMS Microbiol Rev* **31**, 692-720.
Tandeau de Marsac, N. (1977). Occurrence and nature of chromatic adaptation in cyanobacteria. *J Bacteriol* **130**, 82-91.
Tani, T., Gram, L. K., Arakawa, H., Kikuchi, A., Chiba, M., Ishii, Y., Steffansen, B., and Tamai, I. (2008). Involvement of organic anion transporting polypeptide 1a5 (Oatp1a5) in the intestinal absorption of endothelin receptor antagonist in rats. *Pharm Res* **25**, 1085-1091.
Teixera, M., Costa, M., Carvalho, V., Pereira, M., and Hage, E. (1993). Gastroenteritis epidemic in the area of the Itaparica Dam, Bahia, Brazil. *Bulletin of the Pan-American Health Organization* **27**, 244-253.
Teneva, I., Mladenov, R., Popov, N., and Dzhambazov, B. (2005). Cytotoxicity and apoptotic effects of microcystin-LR and anatoxin-a in mouse lymphocytes. *Folia Biol (Praha)* **51**, 62-67.
Tisdale, E. S. (1931). Epidemic of intestinal disorders in Charleston, W. VA., occurring simultaneously with unprecedented water supply conditions. *American Journal of Public Health* **21**, 198-200.

7. References

Toivola, D. M., Eriksson, J. E., and Brautigan, D. L. (1994). Identification of protein phosphatase 2A as the primary target for microcystin-LR in rat liver homogenates. *FEBS Lett* **344**, 175-180.

Toivola, D. M., Goldman, R. D., Garrod, D. R., and Eriksson, J. E. (1997). Protein phosphatases maintain the organization and structural interactions of hepatic keratin intermediate filaments. *Journal of Cell Science* **110**, 23-33.

Toivola, D., Omary, M., Ku, N.-O., Peltola, O., Baribault, H., and Eriksson, J. (1998). Protein phosphatase inhibition in normal and keratin 8/18 assembly-incompetent mouse strains supports a functional role of keratin intermediate filaments in preserving hepatocyte integrity. *Hepatology* **28**, 116-128.

Toivola, D. M., and Eriksson, J. E. (1999). Toxins affecting cell signalling and alteration of cytoskeletal structure. *Toxicology in Vitro* **13**, 521-530.

Tsuji, K., Watanuki, T., Kondo, F., Watanabe, M., Suzuki, S., Nakazawa, H., Suzuki, M., Uchida, H., and Harada, K.-I. (1995). Stability of microcystins from cyanobacteria-II. Effect of UV light on decomposition and isomerization. *Toxicon* **33**, 1619-1631.

Turner, P., Gammie, A., Hollinrake, K., and Codd, G. (1990). Pneumonia associated with contact with cyanobacteria. *British Medical Journal* **300**, 1440-1441.

Ueno, Y., Nagata, S., Tsutsumi, T., Hasegawa, A., Watanabe, M. F., Park, H.-D., Chen, G.-C., Chen, G., and Yu, S.-Z. (1996). Detection of microcystins, a blue-green algal hepatotoxin, in drinking water sampled in Haimen and Fusui, endemic areas of primary liver cancer in China, by highly sensitive immunoassay. *Carcinogenesis* **17**, 1317-1321.

Ueno, Y., Makita, Y., Nagata, S., Tsutsumi, T., Yoshida, F., Tamura, S.-I., Sekijima, M., Tashiro, F., Harada, T., and Yoshida, T. (1999). No chronic oral toxicity of a low dose of microcystin-LR, a cyanobacterial hepatotoxin, in female BALB/c mice. *Environmental Toxicology* **14**, 45-55.

van Apeldoorn, M. E., van Egmond, H. P., Speijers, G. J., and Bakker, G. J. (2007). Toxins of cyanobacteria. *Mol Nutr Food Res* **51**, 7-60.

van den Hoek, C. And Jahns, H. M. (2002). Algae – an introduction tp phycology. University Press. Cambridge.

van der Meer, M. T., Schouten, S., Sinninghe Damste, J. S., de Leeuw, J. W., and Ward, D. M. (2003). Compound-specific isotopic fractionation patterns suggest different carbon metabolisms among Chloroflexus-like bacteria in hot-spring microbial mats. *Appl Environ Microbiol* **69**, 6000-6006.

van Montfoort, J. E., Hagenbuch, B., Groothuis, G. M., Koepsell, H., Meier, P. J., and Meijer, D. K. (2003). Drug uptake systems in liver and kidney. *Current Drug Metabolism* **4**, 185-211.

Vasconcelos, V., Sivonen, K., Evans, W., Carmichael, W., and Namikoshi, M. (1996). Hepatotoxic microcystin diversity in cyanobacterial blooms collected in Portuguese freshwaters. *Water Research* **30**, 2377-2384.

Vasconcelos, V. M. (1999). Cyanobacterial toxins in Portugal: effects on aquatic animals and risk for human health. *Brazilian Journal of Medical and Biological Research* **32**, 249-254.

Vezie, C., Brient, L., Sivonen, K., Bertru, G., Lefeuvre, J., and Salkinoja-Salonen, M. (1998). Variation of Microcystin Content of Cyanobacterial Blooms and Isolated Strains in Lake Grand-Lieu (France). *Microbial Ecology* **35**, 126-135.

Volbracht, C., Leist, M., and Nicotera, P. (1999). ATP controls neuronal apoptosis triggered by microtubule breakdown or potassium deprivation. *Mol Med* **5**, 477-489.

Ward, D., Ferris, M., Nold, S., and Bateson, M. (1998). A natural view of microbial biodiversity within hot spring cyanobacterial mat communities. *Microbiology and Molecular Biology Reviews* **62**, 1353-1370.

Watanabe, M. F., Oishi, S., Harda, K., Matsuura, K., Kawai, H., and Suzuki, M. (1988). Toxins contained in Microcystis species of cyanobacteria (blue-green algae). *Toxicon* **26**, 1017-1025.

Welker, M., and von Dohren, H. (2006). Cyanobacterial peptides - nature's own combinatorial biosynthesis. *FEMS Microbiol Rev* **30**, 530-563.

Weng, D., Lu, Y., Wei, Y., Liu, Y., and Shen, P. (2007). The role of ROS in microcystin-LR-induced hepatocyte apoptosis and liver injury in mice. *Toxicology* **232**, 15-23.

Westhoff, D. E., Rumbley, J. N., Salo, D. R., Rich, T. P., and Anderson, G. W. (2008). Organic anion-transporting polypeptides at the blood-brain and blood-cerebrospinal fluid barriers. *Curr Top Dev Biol* **80**, 135-170.

WHO (1998). Cyanobacterial toxins: Microcystin-LR. In *Guidelines for drinking-water quality*, pp. 95-110. World Health Organization, Geneva.

WHO (1999). *Toxic Cyanobacteria in Water: A guide to their public health consequences, monitoring and management.* F & FN Spon, London.

Wickstrom, M. L., Khan, S. A., Haschek, W. M., Wyman, J. F., Eriksson, J. E., Schaeffer, D. J., and Beasley, V. R. (1995). Alterations in Microtubules, Intermediate Filaments, and Microfilaments Induced by Microcystin-LR in Cultured Cells. *Toxicol Pathol* **23**, 326-337.

Wickstrom, M., Haschek, W., Henningsen, G., Miller, L. A., Wyman, J., and Beasley, V. (1996). Sequential ultrastructural and biochemical changes induced by microcystin-LR in isolated perfused rat livers. *Natural Toxins* **4**, 195-205.

Wiegand, C., and Pflugmacher, S. (2005). Ecotoxicological effects of selected cyanobacterial secondary metabolites: a short review. *Toxicol Appl Pharmacol* **203**, 201-218.

Xie, L., Yokoyama, A., Nakamura, K., and Park, H. (2007). Accumulation of microcystins in various organs of the freshwater snail Sinotaia histrica and three fishes in a temperate lake, the eutrophic Lake Suwa, Japan. *Toxicon* **49**, 646-652.

Xing, M. L., Wang, X. F., and Xu, L. H. (2008). Alteration of proteins expression in apoptotic FL cells induced by MCLR. *Environmental Toxicology* **23**, 451-458.

7. References

Yoon, S., Choi, J., Yoon, J., Huh, J.-W., and Kim, D. (2006). Okadaic acid induces JNK activation, bim overexpression and mitochondrial dysfunction in cultured rat cortical neurons. *Neuroscience Letters* **394**, 190.

Yu, S.-Z. (1989). Drinking water and primary liver cancer. In *Primary liver cancer* (Z. Y. Tang, M. C. Wu, and S. S. Xia, Eds.), pp. 30-37. China Academic Publishers/Springer, New York.

Yu, S.-Z. (1994). Blue-green algae and liver cancer. Toxic Cyanobacteria, Current Status of Research and Management, pp. 75-85.

Yu, S.-Z. (1995). Primary prevention of hepatocellular carcinoma. *Journal of Gastroenterology and Hepatology* **10**, 674-682.

Zaher, H., zu Schwabedissen, H. E., Tirona, R. G., Cox, M. L., Obert, L. A., Agrawal, N., Palandra, J., Stock, J. L., Kim, R. B., and Ware, J. A. (2008). Targeted disruption of murine organic anion-transporting polypeptide 1b2 (Oatp1b2/Slco1b2) significantly alters disposition of prototypical drug substrates pravastatin and rifampin. *Mol Pharmacol* **74**, 320-329.

Zegura, B., Sedmak, B., and Filipic, M. (2003). Microcystin-LR induces oxidative DNA damage in human hepatoma cell line HepG2. *Toxicon* **41**, 41-48.

Zegura, B., Lah, T. T., and Filipic, M. (2004). The role of reactive oxygen species in microcystin-LR-induced DNA damage. *Toxicology* **200**, 59-68.

Zegura, B., Zajc, I., Lah, T. T., and Filipic, M. (2008). Patterns of microcystin-LR induced alteration of the expression of genes involved in response to DNA damage and apoptosis. *Toxicon* **51**, 615-623.

Zhang, Q.-X., Carmichael, W. W., Yu, M.-J., and Li, S.-H. (1991). Cyclic peptide hepatotoxins from freshwater cyanobacterial (blue-green algae) waterblooms collected in Central China. *Environmental Toxicology and Chemistry* **10**, 313-321.

Zhou, L., Yu, D., and Yiu, H. (2000). Drinking water types, microcystins and colorectal cancer. *Zhonghua-Yufang-Yixue-Zazhi* **34**, 224-226.

Zhou, L., Yu, H., and Chen, K. (2002). Relationship between microcystin in drinking water and colorectal cancer. *Biomedical & Environmental Sciences* **15**, 166-171.

Zilberg, B. (1966). Gastroenteritis in Salisbury European children-a five-year study. *The Central African Journal of Medicine* **12**, 164-168.

Zurawell, R. W., Chen, H., Burke, J. M., and Prepas, E. E. (2005). Hepatotoxic cyanobacteria: a review of the biological importance of microcystins in freshwater environments. *J Toxicol Environ Health B Crit Rev* **8**, 1-37.

Appendices

Abbreviations

AD	Alzheimer's Disease
Adda	3-amino-9-methoxy-2,6,8-trimethyl-10-phenyldeca-4,6-dienoic acid
AFA	*Aphanizomenon flos-aquae*
BBB	blood-brain-barrier
BCSFB	blood-cerebrospinal-fluid-barrier
BGAS	blue-green algae supplements
BMAA	β-methyl-amino-L-alanine
Bq	Becquerel
BSP	bromosulfopthalein
bw	body weight
cPPIA	colorimetric protein phosphatase inhibition assay
Da	Dalton
DEVD-AFC	asp-glu-val-asp-7-amino-4-trifluormethylcoumarin
DTT	dithiothreitol
dw	dry weight
ES	estrone sulfate
GAPDH	glyceraldehyde-3-phosphate-dehydrogenase
^{3}[H]	indicates a tritiated substrate
IRTG 1331	International Research Training Group 1331
i.p.	intraperitoneal
i.v.	intravenous
MTT	3-(4,5 dimethylthiazol-2-yl)-2,5-diphenyl tetrazolium bromide
MC	microcystin
MC$_{equiv}$	microcystin equivalents
MC-AR	microcystins including alanine and arginine
MC-LR	microcystins including leucine and arginine
MC-LW	microcystins including leucine and tryptophan
MC-LF	microcystins including leucine and phenylalanine
MC-YR	microcystins including tyrosine and arginine
mCGC	murine cerebellar granule cells
mOatp	murine organic anion transporting polypeptide

mWBC	murine whole brain cells
nm	nanometer
NOEL	non abserved effect level
OA	okadaic acid
Oatp	rodent organic anion transporting polypeptide
OATP	human organic anion transporting polypeptide
PFA	paraformaldehyd
p.o.	per oral
PP	protein phosphatase
PSP	paralytic shellfish poisoning
RT	room temperature
SDS-PAGE	sodium dodecyl sulfate-polyacrylamide gel electrophoresis
ser	serine
Stsp	staurosporine
STX	saxitoxin
TC	taurocholate
TDI	tolerable daily intake
thr	threonine
UV	ultraviolet
WHO	World Health Organisastion
WB	western-blot

Danksagung

Zum Schluß möchte ich mich noch bei einigen Menschen herzlich bedanken, die mich in unterschiedlichster Weise während der letzten drei Jahre begleitet, unterstützt und zum Gelingen dieser Doktorarbeit beigetragen haben.

Mein ganz besonderer Dank gilt:

Prof. Dr. Daniel R. Dietrich für dieses hochspannende Dissertationsthema, der Möglichkeit, dieses Projekt als Kollegiat der International Resaerch Training Group 1331 durchzuführen, sein Vertrauen in mich sowie die stetige wissenschaftliche und menschliche Unterstützung.

Priv. Doz. Dr. Elisa May für die Übernahme des Koreferates, das Interesse an dieser Arbeit und ihre Unterstützung bei unterschiedlichsten mikroskopischen und confocal mikroskopischen Anwendungen.

Prof. Dr. Klaus P. Schäfer und Prof. Dr. Bruno Stieger, als IRTG 1331 PhD thesis commitee, für ihre konstruktiven Beiträge, die maßgeblich zum Gelingen dieser Doktorarbeit beigetragen haben.

Prof. Dr. Albrecht Wendel als ehemaliger- und Prof. Dr. Marcel Leist als derzeitiger Vorsitzende des IRTG 1331 für die Aufnahme in das Kolleg sowie die Finanzierung von Projekten, Fortbildungskursen und Kongressen.

Meinen IRTG 1331 Kollegen für zahlreiche wissenschaftliche Beiträge sowie soziale und kulturelle Erlebnisse während Meetings und Fortbildungskursen im In- und Ausland, besonders zu erwähnen Tobias Speicher, Stephan Penzkofer, Christine Strasser, Diana Scholz, Christine Hoffmann und Stephanie Siegl.

Dr. Valerié Fessard, Dr. Ludovic Le Hegarat and collegues from AFSSA, Javené, France, for their readiness to collaborate with us, their great support during my stays, as well as for the nice evenings with excellent french food.

Dr. Bernhard Ernst und Andreas Fischer für zahlreiche Hilfestellungen, deren Diskussionsbereitschaft sowie die schönen Abende in Brasilien.

Kerstin, Alex, Heiko und Julia für theoretische und praktische Unterstützungen, die wesentlich zum Fortschritt dieser Doktorarbeit beigetragen haben. Herzliches Dankeschön gilt Andrew Borkowsky für die Englisch-Korrekturen.

Meiner Familie und Nicole, die mich seit Jahren tatkräftig unterstützen, mir Rückhalt geben und somit das Fundament dieser Arbeit geschaffen haben.

I want morebooks!

Buy your books fast and straightforward online - at one of world's fastest growing online book stores! Environmentally sound due to Print-on-Demand technologies.

Buy your books online at
www.morebooks.shop

Kaufen Sie Ihre Bücher schnell und unkompliziert online – auf einer der am schnellsten wachsenden Buchhandelsplattformen weltweit! Dank Print-On-Demand umwelt- und ressourcenschonend produziert.

Bücher schneller online kaufen
www.morebooks.shop

KS OmniScriptum Publishing
Brivibas gatve 197
LV-1039 Riga, Latvia
Telefax: +371 686 204 55

info@omniscriptum.com
www.omniscriptum.com

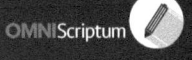

Printed by Books on Demand GmbH, Norderstedt / Germany